Universe Beyond Imagination - Game of Time

Debasish Talukdar

Published by Deb T, 2023.

While every precaution has been taken in the preparation of this book, the publisher assumes no responsibility for errors or omissions, or for damages resulting from the use of the information contained herein.

UNIVERSE BEYOND IMAGINATION - GAME OF TIME

First edition. April 5, 2023.

Copyright © 2023 Debasish Talukdar.

ISBN: 979-8201103255

Written by Debasish Talukdar.

Table of Contents

To all those curious minds who love physics

GAME OF TIME

INTRODUCTION

Time is a fundamental concept that has captivated humans for centuries. It is an ever-present force in our lives, and yet it remains a mystery that we are still trying to understand. We measure time, we plan our lives around it, and we experience it in countless ways, but what exactly is time?

In this book, we will delve into the nature of time, exploring its philosophical, physical, and cultural dimensions. We will examine how our understanding of time has evolved over time and how it has shaped our perception of the world around us. We will also explore some of the most fascinating and perplexing questions related to time, including whether time travel is possible, and whether time is a dimension.

Throughout this book, we will explore the many facets of time, from its role in physics and astrophysics to its impact on human psychology and behaviour. We will also examine the cultural significance of time, and how different societies have conceptualized and measured it throughout history.

As we explore these different dimensions of time, we will gain a deeper appreciation for this enigmatic concept and the many ways in which it shapes our world and our lives.

Definition of time

Time is one of the most fundamental and universal concepts that we use to make sense of the world around us. We use it to understand our past, our present, and to plan for the future. We organize our lives around the concept of time, from the ticking of a clock to the seasonal changes in nature.

But what exactly is time? How do we define it, and what makes it such a crucial concept in our lives? In this chapter, we will explore the various definitions of time that have been proposed throughout history and examine the different ways in which we can understand this fundamental concept.

At its most basic level, time can be defined as a measure of duration or the interval between two events. This definition is often represented by the ticking of a clock or the movement of a celestial object, such as the rotation of the Earth or the orbit of the moon around the Earth. In this sense, time is seen as a physical phenomenon that can be measured and quantified.

However, this simple definition does not fully capture the complexity of time. Time is not just a measure of duration, but it is also a fundamental aspect of the universe. It affects the way we perceive reality, the way we understand causality and change, and the way we interact with the world around us.

Philosophers, scientists, and scholars have offered different definitions of time throughout history, often reflecting the cultural and historical contexts in which they were working. In ancient Greece, time was seen as cyclical and infinite, while in medieval Europe, time was seen as a linear progression leading towards a final judgment.

In the modern era, time has been defined in many different ways, often reflecting the advancements in science and technology. For example, in the theory of relativity, time is seen as a dimension that is intimately linked to space. In this view, time is not just a measure of duration but is a fundamental aspect of the fabric of the universe.

However, not all definitions of time are based on a physical understanding. Some philosophers and scholars see time as a subjective experience that is shaped by our individual perceptions and experiences. In this view, time is not just an external phenomenon that can be measured but is a deeply personal and subjective concept.

While it can be seen as a physical phenomenon that can be measured and quantified, it is also a fundamental aspect of the universe that affects the way we perceive reality and interact with the world around us. Ultimately, the definition of time is shaped by our cultural and historical context and reflects the ongoing evolution of our understanding of the world.

Despite the different definitions of time, one thing is clear: it is an essential concept that plays a central role in our lives. We use time to structure our daily routines, plan for the future, and make sense of our past experiences. It is a concept that is so deeply ingrained in our lives that it can be difficult to imagine a world without it. Scientists and philosophers continue to grapple with the nature of time and its relationship with other fundamental concepts such as space and causality.

One of the most significant challenges in understanding time is its apparent directionality. We experience time as moving in one direction only, from past to present to future. This has led some scholars to propose that time is not just a physical phenomenon but is also intimately linked to our experience of consciousness.

In recent years, advancement in physics have opened up new avenues for understanding time. The theory of relativity, for example, has shown that time is not a fixed and immutable concept but can be influenced by the properties of space and matter. The concept of spacetime, which unifies space and time into a single four-dimensional continuum, has led to new insights into the nature of time and its relationship to the physical world.

Despite these advancements, there is still much that we do not understand about time. The possibility of time travel, for example, remains a subject of much debate and speculation. While some scientists and philosophers believe that time travel may be possible in theory, others argue that it is simply a product of science fiction and is unlikely to ever become a reality.

While it can be seen as a physical phenomenon that can be measured and quantified, it is also a fundamental aspect of our consciousness and subjective experience. Despite our ongoing efforts to understand it, time remains one of the most enigmatic and mysterious concepts in the universe, and it is likely to continue to challenge and inspire us for generations to come.

Importance of Studying Time

Time is a fundamental concept that underpins many areas of human inquiry and endeavour. From philosophy and physics to history and literature, time has been a central focus of human inquiry for thousands of years. In this chapter, we will explore some of the key reasons why the study of time is so important.

Understanding the nature of time

Perhaps the most obvious reason for studying time is to gain a deeper understanding of the nature of the concept itself. What is time? How does it relate to other fundamental concepts such as space and causality? These are questions that have intrigued scientists and philosophers for centuries. Through the study of time, we can gain a better understanding of the physical world and the laws that govern it. For example, the theory of relativity has shown that time is not an absolute and unchanging concept but can be influenced by the properties of space and matter. By studying time, we can gain a deeper appreciation for the complexities and subtleties of the physical universe.

Making sense of history

The study of time is also essential for making sense of history. Understanding the sequence of events and their relationship to each other is critical for understanding the past and how it has shaped the present. By studying the ways in which people have thought about and experienced time throughout history, we can gain valuable insights into the cultural and social context of different periods. For example, the concept of time has played a critical role in shaping religious beliefs, scientific discoveries, and political ideologies.

Planning for the future

Time is also essential for planning and preparing for the future. From setting goals and making long-term plans to managing our daily schedules, time is a critical component of our everyday lives. By studying time, we can develop a better understanding of how to manage our time effectively, make the most of our resources, and achieve our goals. This is particularly important in today's fast-paced and increasingly complex world, where time management skills are essential for success in both personal and professional contexts.

Advancing scientific understanding

Finally, the study of time is crucial for advancing our understanding of the universe and the laws that govern it. From the mysteries of black holes and the nature of the Big Bang to the fundamental properties of matter and energy, time plays a central role in most of the significant scientific questions of our time. Through the study of time, scientists can develop new theories, conduct ground-breaking experiments, and make discoveries that have the potential to transform our understanding of the universe and our place within it.

Implications for personal growth and well-being

Beyond its scientific and historical importance, the study of time also has significant implications for personal growth and well-being. For example, research has shown that people who feel like they have more control over their time tend to experience greater levels of happiness and life satisfaction. By studying time and how we use it, we can develop a better understanding of our priorities, values, and goals. This can help us in making more intentional choices about how we spend our time, leading to greater fulfillment and satisfaction in life.

Cultural and societal implications

Finally, the study of time has important cultural and societal implications. Different cultures have different attitudes and beliefs about time, which can shape everything from work and leisure to social norms and values. By studying time in different cultural and societal contexts, we can gain a better understanding of how time is perceived and valued in different parts of the world. This can help us develop a more inclusive and nuanced understanding of the ways in which time shapes our lives and societies.

The study of time is a multifaceted and complex field that has significant implications for a wide range of disciplines and areas of inquiry. From understanding the nature of the physical universe to improving our personal well-being and advancing scientific understanding, time is a fundamental concept that plays a critical role in many aspects of human life. By studying time, we can gain new insights and perspectives that have the potential to transform the world around us. The study of time is essential for gaining a deeper understanding of the world around us. From understanding the nature of time, itself to making sense of history and planning for the future, time plays a critical role in many areas of human inquiry and endeavour. By studying time, we can develop new insights and discoveries that have the potential to shape our understanding of the universe for generations to come.

THE NATURE OF TIME
Historical Perspectives on Time

The concept of time has been explored and debated by philosophers, scientists, and scholars for centuries. In this chapter, we will examine some of the key historical perspectives on time, tracing the evolution of our understanding of this fundamental concept over time.

Ancient Perspectives:

In many ancient cultures, time was viewed as cyclical and repetitive, with events and phenomena repeating themselves in a never-ending cycle. The Babylonians, for example, developed a sophisticated system of astronomy that allowed them to predict celestial events such as eclipses and solstices with remarkable accuracy.

In ancient Greece, philosophers such as Aristotle and Plato viewed time as an abstract and immutable concept that existed independently of human experience. Aristotle famously defined time as "a number of motions in respect of before and after," suggesting that time was closely linked to the concept of change.

In ancient times, the concept of time was closely tied to the cycles of nature, such as the phases of the moon and the changing of the seasons. The Babylonians, for example, used a lunar calendar to keep track of time, while the Egyptians based their calendar on the annual flooding of the Nile River.

The ancient Greeks also had a unique perspective on time. They saw time as a process of change, and they believed that everything in the universe was subject to this process. They also saw time as being cyclical, with events repeating themselves in a never-ending cycle.

The ancient Romans, on the other hand, saw time as a linear progression, with events unfolding in a predetermined sequence. They were also the first to use the Julian calendar, which was later refined by the Gregorian calendar.

In ancient China, time was viewed as a combination of cyclical and linear elements. The Chinese lunar calendar was based on the cycles of the moon, while their solar calendar was based on the annual movements of the sun.

The Mayan civilization also had a complex understanding of time, with their calendar system consisting of multiple cycles that tracked the movements of the planets and stars. They believed that time was cyclical and that events repeated themselves in cycles.

The ancient perspective on time was heavily influenced by the natural world and its cycles. Time was seen as an essential part of the universe, and its cyclical and linear elements were intricately woven together. Understanding these ancient perspectives can provide us with a deeper appreciation for the way in which our ancestors viewed the world and the role that time played in their lives.

The ancient perspective on time was not only limited to the cyclical and linear elements, but it also had a significant influence on religious and philosophical beliefs. For example, in ancient Hindu philosophy, time was viewed as a manifestation of the divine, and it was considered to be eternal and infinite. The concept of karma, which is the idea that our actions in this life determine our fate in the next, was closely tied to the Hindu concept of time.

Similarly, in ancient Greek philosophy, time was considered to be intimately connected to the concept of change, which was seen as a fundamental aspect of existence. The philosopher Heraclitus famously stated that "no man ever steps in the same river twice, for it's not the same river and he's not the same man." This idea reflects the Greek belief that time is constantly moving forward, and that nothing remains the same.

The ancient perspective on time also had a significant impact on the development of science and technology. For example, the Babylonians were able to predict the movements of the planets and stars using their intricate astronomical knowledge, which was based on their understanding of the cycles of nature.

The ancient Greeks made significant contributions to the field of mathematics, which helped to establish the foundations of modern science. The Greek mathematician Euclid, for example, wrote a book called "Elements," which is still considered to be one of the most influential mathematical texts of all time.

Time was seen as an essential part of the universe, and its cyclical and linear elements were closely intertwined with religious, philosophical, and scientific beliefs. By understanding these ancient perspectives, we can gain a deeper appreciation for the way in which our understanding of time has evolved over time.

Medieval Perspectives:

During the medieval period, time was often viewed as a manifestation of God's will, with the universe unfolding according to a divine plan. This view was heavily influenced by the teachings of the Christian church, which emphasized the importance of divine intervention in the affairs of the world. During the Middle Ages, the Christian Church held significant influence over the way in which people viewed time. The medieval perspective on time was heavily influenced by religious beliefs, and time was seen as a reflection of God's plan for the universe.

The Church viewed time as being divided into two distinct parts: the time before the coming of Christ (known as "BC" or "Before Christ"), and the time after his birth (known as "AD" or "Anno Domini," which means "in the year of our Lord"). This division of time was significant because it marked the beginning of a new era in human history.

One of the most significant contributions of the medieval period to our understanding of time was the development of the mechanical clock. Prior to the invention of the mechanical clock, time was measured using natural phenomena such as the movement of the sun and stars. However, the mechanical clock allowed people to measure time more accurately, which was

essential for the development of science and technology.

The medieval perspective on time was also closely tied to the concept of fate. Many people believed that their lives were predetermined by God, and that events would unfold according to his plan. This belief in fate was often reflected in works of literature, such as the epic poem "Beowulf," which features a hero who is fated to die in battle.

The medieval period also saw the development of the concept of linear time, which was influenced by the teachings of the ancient Greek philosopher Aristotle. According to Aristotle, time is a continuous and uniform progression from the past to the present and into the future. This concept of linear time was significant because it helped to lay the foundations for modern science, which relies on the idea that natural phenomena follow consistent patterns over time.

The medieval perspective on time was heavily influenced by religious beliefs, and time was seen as a reflection of God's plan for the universe. The development of the mechanical clock and the concept of linear time were significant contributions of the medieval period to our understanding of time. By understanding these medieval perspectives, we can gain a deeper appreciation for the way in which our understanding of time has been shaped by religious, philosophical, and scientific beliefs throughout history.

In addition to the medieval perspectives on time shaped by religious beliefs, there were also significant contributions from scholars and philosophers during this period. One of the most influential philosophers of the medieval period was Saint Augustine, who wrote extensively on the nature of time and its relationship to God.

Augustine's perspective on time was heavily influenced by his Christian beliefs. He argued that time was a creation of God and that it was a necessary component of the universe. According to Augustine, time is both subjective and objective. Subjectively, time is the measure of the movement of the mind, while objectively, time is the measure of the movement of physical objects.

Augustine also explored the concept of eternity, which he viewed as a timeless state of being outside of the physical world. He argued that eternity was the ultimate goal of human existence, and that it could be achieved through a connection with God.

Another significant figure in medieval philosophy was Thomas Aquinas, who built upon Augustine's ideas and expanded upon them in his own work. Aquinas believed that time was a continuous flow that was essential for the existence of the universe. He also believed that time was a measure of change and that change was a necessary component of the physical world.

Aquinas also explored the concept of timelessness, which he viewed as a state of being outside of time that could be achieved through a connection with God. He argued that timelessness was the ultimate goal of human existence and that it could be achieved through contemplation and spiritual practices.

The medieval period was a time of significant development in our understanding of time. The perspectives on time during this period were heavily influenced by religious beliefs, but they also drew upon the ideas of philosophers and scholars. Through the work of figures such as Augustine and Aquinas, the medieval period contributed important ideas to our understanding of time that continue to shape our thinking today.

In addition to the perspectives on time put forth by religious figures and philosophers, the medieval period also saw advancements in the measurement of time. The development of mechanical clocks, for example, allowed for more precise measurements of time and facilitated the standardization of time across different locations.

One of the most significant advances in time measurement during the medieval period was the invention of the hourglass. Hourglasses, also known as sandglasses, were used to measure time by allowing sand to flow through a narrow opening from one chamber to another. By measuring the amount of sand that had passed through the hourglass, people were able to estimate the amount of time that had elapsed.

The use of hourglasses had a significant impact on society, particularly in the fields of navigation and astronomy. The ability to measure time more accurately and consistently allowed for more precise calculations of the positions of the stars and planets, which in turn facilitated advancements in navigation and exploration.

In addition to the hourglass, other mechanical devices such as water clocks and candle clocks were also used during the medieval period to measure time. These devices were not as precise as modern clocks and watches, but they represented important steps forward in our ability to measure and understand the nature of time.

The medieval period was a time of significant progress in our understanding and measurement of time. The perspectives on time put forth during this period were heavily influenced by religious beliefs and philosophical ideas, but they were also shaped by practical concerns such as the need for more precise timekeeping. The advancements made during this period set the stage for further developments in time measurement and understanding that continue to this day.

Another significant contribution to the medieval perspective on time came from the work of the Islamic scholar al-Biruni. In his book "The Masudic Canon," written in the 11th century, al-Biruni proposed a theory of time that suggested that time was continuous and infinite, and that it could not be divided into discrete units or moments.

This idea was quite different from the prevailing Aristotelian view of time, which held that time was made up of discrete, indivisible units called "instants." Al-Biruni's theory of time helped to challenge and expand the medieval understanding of time, paving the way for new and more nuanced perspectives on this fundamental aspect of our existence.

The medieval period also saw the development of a variety of timekeeping devices that were used for practical and ceremonial purposes. One example is the use of church bells to mark the passage of time and signal the start of religious services. This tradition of using bells to mark time continues to this day, and has become an important part of many cultures around the world.

The medieval perspective on time was shaped by a complex interplay of philosophical, religious, and practical concerns. The advancements made during this period paved the way for further progress in time measurement and understanding, and helped to lay the foundation for the modern scientific view of time that we have today.

MODERN PERSPECTIVES:

In the modern era, the scientific understanding of time has been greatly influenced by the theories of relativity and quantum mechanics. These theories have fundamentally challenged our traditional understanding of time as a universal and absolute concept.

Albert Einstein's theory of relativity proposes that time is relative, meaning it can vary depending on an observer's relative motion and the presence of gravity. According to relativity, time can be seen as a fourth dimension, and the concept of time dilation suggests that time passes more slowly in areas of strong gravitational forces or when moving at high speeds.

Quantum mechanics has also challenged our understanding of time. In the quantum world, particles do not have a fixed position or momentum until they are observed, and time seems to operate differently than it does in classical physics. The concept of entanglement, where two particles can be linked regardless of distance, has also raised questions about the nature of time and causality.

These scientific discoveries have also had practical applications, such as the use of atomic clocks for precise timekeeping and the development of GPS technology. Furthermore, the study of time has expanded beyond physics to fields such as psychology, sociology, and philosophy.

In the modern era, time has become increasingly interconnected with technology and the global economy. The concept of time zones and standard time has allowed for efficient communication and travel across vast distances. However, this interconnection has also led to a culture of constant connectivity, with people feeling the pressure to be always available and productive.

Overall, the modern perspective on time is one that recognizes its relativity and complexity, as well as its crucial role in both the natural world and human society.

One of the most significant developments in the modern perspective of time is the understanding of time as a resource. In our fast-paced society, time is seen as a limited resource that must be managed and maximized for productivity and success. This has led to a culture of time management techniques and tools, such as calendars, to-do lists, and productivity apps, to help individuals make the most of their time.

The study of time has also expanded beyond the physical sciences to fields such as psychology, sociology, and philosophy. In psychology, the study of time perception and time management has shed light on how individuals perceive and utilize time in their daily lives. In sociology, the concept of social time has been used to understand how social structures and culture influence our perception and use of time.

Philosophy has long grappled with the concept of time, with philosophers such as Immanuel Kant and Henri Bergson proposing different theories about the nature of time. Kant saw time as a necessary structure of the human mind, while Bergson proposed a view of time as a continuous flow or duration.

The modern perspective on time is also shaped by cultural and societal values. For example, punctuality and efficiency are highly valued in Western cultures, while in some other cultures, such as in the Middle East and Latin America, a more relaxed approach to time is often taken. The concept of time is also shaped by religion, with different faiths having varying views on the nature of time and its role in the universe.

The study of time has expanded beyond the physical sciences to encompass a wide range of fields and perspectives, highlighting its crucial role in shaping human society and culture.

In addition to the understanding of time as a resource and its role in human society, the modern perspective on time has also been shaped by scientific advancements. The theory of relativity, proposed by Albert Einstein in the early 20th century, revolutionized our understanding of time and its relationship to space.

Einstein's theory proposed that time is not absolute and that the perception of time is relative to the observer's motion and position. This led to the concept of time dilation, where time appears to move slower in a higher gravitational field or when an object is moving at high speeds.

The theory of relativity also brought about the concept of spacetime, where time and space are seen as interlinked dimensions. This has led to new theories and speculations about time travel and the possibility of manipulating time through space-time distortion.

Another important development in the modern perspective of time is the study of the origins of the universe and the nature of time at the beginning of time itself. The Big Bang theory, which proposes that the universe began as a singularity and has been expanding ever since, has led to new questions about the nature of time and the concept of time before the Big Bang.

The modern perspective on time is one that is constantly evolving and expanding, as new scientific discoveries and cultural shifts continue to shape our understanding of this complex and fundamental concept.

Contemporary Perspectives:

In the contemporary perspective, the study of time has expanded beyond the traditional domains of philosophy and physics to include a range of interdisciplinary fields such as psychology, sociology, anthropology, and neuroscience.

In psychology, the study of time perception has become an important area of research, exploring how the brain processes and experiences the passage of time. Studies have shown that time perception can be influenced by a variety of factors, such as attention, emotion, and context, and that individuals can have differing subjective experiences of time.

Sociologists and anthropologists have also explored the cultural and social dimensions of time, examining how different cultures and societies organize and value time, and how these values impact individual and collective behaviour. For example, the concept of "time orientation" has been used to describe how different cultures prioritize past, present, or future time perspectives.

Neuroscience research has also shed light on the neurological mechanisms underlying time perception and time-related cognitive functions, such as memory and decision-making. Studies have shown that different regions of the brain are involved in processing different aspects of time, such as duration and timing, and that disruptions to these regions can lead to deficits in time perception and related cognitive functions.

Contemporary perspectives on time also continue to be shaped by technological advancements and cultural shifts. The increasing prevalence of digital technologies and the internet has transformed the way we experience and interact with time, leading to new questions and debates about the impact of these technologies on our sense of time and attention.

In addition to the interdisciplinary nature of contemporary perspectives on time, there are also ongoing debates and discussions about the nature and implications of time in various domains.

In physics, for example, the concept of time is still the subject of much debate and discussion. The theory of relativity has challenged traditional notions of time as a constant and absolute entity, proposing instead that time is relative to the observer and that time and space are intimately interconnected. This has led to fascinating discussions about the nature of time, such as whether time is a physical entity or simply a mental construct, and whether time travel is possible.

In philosophy, the concept of time continues to be a rich area of study and debate. Philosophers have explored questions such as the nature of time itself, the relationship between time and causality, and the implications of time for free will and determinism.

Overall, the contemporary perspective on time is one that reflects the growing recognition of the importance and complexity of time across a range of fields and domains. It is a perspective that is characterized by ongoing exploration, debate, and discovery, and that promises to yield new insights and understanding about this fundamental and fascinating aspect of our experience.

How we measure time?

Time is an abstract concept that we use to measure the duration of events and processes. But how do we actually measure time? Let us explore the various methods and devices that humans have used throughout history to measure time, from ancient sundials to modern atomic clocks.

A. Sundials

One of the oldest and simplest ways to measure time is through the use of a sundial. A sundial is essentially a flat plate with a raised pointer called a gnomon, which casts a shadow onto the plate. As the sun moves across the sky, the shadow moves along the plate, indicating the time of day. Sundials have been used for thousands of years, with some of the earliest examples dating back to ancient Egypt.

B. Water Clocks

Water clocks, also known as clepsydras, are another early method of measuring time. These clocks use the flow of water to measure time. A container with a small hole at the bottom is filled with water, and as the water flows out through the hole, it marks the passage of time. Water clocks have been used in various forms since ancient times, and were particularly popular in ancient Greece and Rome.

C. Mechanical Clocks

Mechanical clocks were first developed in Europe in the 14th century. These clocks use a series of gears and weights to keep time. The earliest mechanical clocks were large and expensive, and were mainly used in churches and other public places. Over time, however, mechanical clocks became smaller and more affordable, and eventually became a common household item.

D. *Quartz Clocks*

Quartz clocks use a piece of quartz crystal to keep time. When an electrical current is applied to the crystal, it vibrates at a very specific frequency, which is used to keep time. Quartz clocks were first developed in the 1920s, and quickly became popular due to their accuracy and affordability. Today, most household clocks and watches use quartz technology.

E. *Atomic Clocks*

Atomic clocks are the most accurate clocks in existence, and are used to keep time for scientific research, navigation, and other purposes that require extreme precision. These clocks use the vibrations of atoms to keep time. The most common type of atomic clock is the caesium atomic clock, which measures the vibrations of caesium atoms to keep time.

Humans have been measuring time for thousands of years, using a variety of methods and devices. From sundials to atomic clocks, our methods for measuring time have become increasingly accurate and precise over time, reflecting our ongoing fascination with this fundamental aspect of our existence.

Einstein's theory of relativity and the concept of spacetime

Albert Einstein's theory of relativity revolutionized our understanding of space and time. The theory has two parts, the special theory of relativity and the general theory of relativity. The special theory of relativity deals with objects moving in a straight line at a constant velocity relative to one another, while the general theory of relativity deals with objects in the presence of gravity.

One of the central ideas of Einstein's theory of relativity is that space and time are not separate and distinct entities, but are instead combined into a single entity called spacetime. In this view, time is not absolute, but is instead relative to the observer's reference frame. This means that time can appear to pass more slowly or quickly depending on how fast an observer is moving and how strong the gravitational field they are in.

Einstein's theory also introduced the concept of time dilation, which means that time appears to pass more slowly for objects that are moving at high speeds relative to a stationary observer. This effect has been observed in many experiments, including the famous "twin paradox," where one twin travels at high-speed relative to the other twin, causing the traveling twin to experience less time than the stationary twin.

The concept of spacetime also led to the idea of a "light cone," which is a diagram that represents the possible paths of light from an event in spacetime. The light cone divides spacetime into two regions: the future light cone and the past light cone. The future light cone represents events that can be affected by an event, while the past light cone represents events that can affect the event.

Einstein's theory of relativity has had a profound impact on our understanding of the universe.It has led to new discoveries and technologies, such as the Global Positioning System (GPS), which relies on the fact that time is affected by gravity and motion. It has also challenged our intuitive understanding of space and time and has shown that our experiences of the world are not always what they seem.

Einstein's theory of relativity and the concept of spacetime have fundamentally changed our understanding of time and space. The theory has shown that time is not absolute, but is instead relative to the observer's reference frame. This has led to new discoveries and technologies and has challenged our intuitive understanding of the universe.

Now what is twin paradox?

The twin paradox is a famous thought experiment that is often used to explain the implications of Einstein's theory of relativity on time. It is also known as the clock paradox or the twin "paradox" because it seems to contradict our everyday understanding of time.

The thought experiment involves two identical twins, one of whom travels to a distant planet at near the speed of light while the other remains on Earth. According to the theory of relativity, time passes more slowly for an object in motion relative to an observer at rest. Therefore, the twin on the spaceship will experience time passing more slowly than the twin on Earth.

When the traveling twin returns to Earth after some time, they will have aged less than their twin who stayed on Earth. This is because the time dilation effect caused by their high-speed travel slowed down the rate at which time passed for them.

The twin paradox is paradoxical because it seems to suggest that both twins can be younger than the other. However, this is only true in the context of relativity, where time is not an absolute quantity and can be affected by the relative motion of an observer.

The twin paradox has been confirmed experimentally by measuring the decay rate of subatomic particles traveling at high speeds, which slows down relative to an observer at rest. The implications of the twin paradox and relativity have had a profound impact on our understanding of the nature of time and have led to significant advancements in fields such as particle physics and astronomy.

THE PHILOSOPHY OF TIME

Philosophical debates about the nature of time:

Time has been a subject of debate in philosophy for centuries. Philosophers have long debated the nature of time, including its existence, the relationship between time and change, and the possibility of time travel. In this chapter, we will explore some of the major philosophical debates surrounding time.

A. The existence of time

The existence of time is a topic of much philosophical debate. One of the main questions is whether time is a real, objective feature of the universe or simply a subjective experience in the mind.

Those who argue for the objective existence of time point to the fact that events seem to happen one after another in a sequence, and that this sequence is consistent across different observers. They also argue that the laws of physics seem to be time-dependent, and that time is necessary for causation to occur.

On the other hand, those who argue that time is a subjective experience point out that our perception of time can vary depending on our state of mind, and that time can appear to be relative depending on our frame of reference. They also argue that time seems to be a product of the mind's attempt to make sense of the world around us, rather than a fundamental aspect of the universe.

One notable philosopher who has written extensively about the nature of time is Henri Bergson. Bergson argued that time is not a linear sequence of discrete moments, but rather a continuous and flowing process that cannot be captured by the static measurements of a clock. He believed that the human experience of time is subjective and that the mind constructs time as a means of making sense of the world around us.

Another philosopher who has contributed to the debate on the existence of time is J.M.E. McTaggart. McTaggart argued that time is not an objective feature of the universe, but rather an illusion created by the limitations of our consciousness. He claimed that the idea of past, present, and future is self-contradictory, and that time is a logical impossibility.

Despite these conflicting viewpoints, the nature of time remains a fundamental question in philosophy. It is a topic that continues to inspire debate and discussion among thinkers of all disciplines, and it is likely that the philosophical exploration of time will continue for many years to come.

The debate about the existence of time has been a topic of interest for many philosophers, physicists, and scientists for centuries. On one hand, some argue that time is an objective reality that exists independently of our perception, while others claim that time is a subjective construct created by our consciousness.

One of the most influential arguments against the existence of time was put forth by the philosopher J.M.E. McTaggart in his paper "The Unreality of Time" in 1908. McTaggart argued that time cannot exist because it is contradictory and cannot be adequately described by either the A-series or B-series theories of time.

The A-series theory of time suggests that time is a series of events that occur in a linear sequence, with past, present, and future all having distinct objective realities. However, McTaggart argued that this theory was flawed because the concept of time as a present moment is inherently contradictory. He believed that the present is constantly shifting and can never be defined as an objective reality.

The B-series theory of time, on the other hand, suggests that time is a series of events ordered in a logical sequence, with past, present, and future all existing simultaneously as objective realities. However,

McTaggart argued that this theory was also flawed because it was unable to account for the subjective experience of time and the feeling of the present moment.

Other philosophers have challenged McTaggart's argument and have presented their own theories about the existence of time. For example, the philosopher Gottfried Wilhelm Leibniz argued that time is a relative concept and that the universe is made up of an infinite number of "monads," each with its own individual time frame.

In physics, the concept of time is essential to our understanding of the universe. The theory of relativity suggests that time is relative and can be influenced by factors such as gravity and velocity. This means that time can appear to move more slowly or quickly depending on the observer's position and motion.

Despite the ongoing philosophical debates about the existence of time, it is clear that time plays a crucial role in our lives and our understanding of the world around us. Whether time is an objective reality or a subjective construct, it remains a fundamental aspect of human existence and experience.

B. THE RELATIONSHIP between time and change

The relationship between time and change has been a topic of philosophical inquiry for centuries. It has been debated whether time is the cause of change or if change is the cause of time. Some philosophers argue that change is the fundamental concept, and time is simply the measurement of change. Others believe that time exists independently of change and that change is simply an event that occurs within the framework of time.

One prominent view is that time is intimately connected with the concept of causation. This view argues that time is necessary for causation to occur, as causation involves a temporal sequence of events. For example, if event A causes event B, then A must occur before B.

This temporal ordering is essential for causation to make sense, and thus, time is necessary for causation to occur.

Another philosophical debate regarding time is whether it is an objective or subjective phenomenon. Objectivists argue that time exists independently of human perception, and its passage is a physical reality that can be measured and studied objectively. Subjectivists, on the other hand, argue that time is a subjective experience that is dependent on the observer's perception and cannot be measured objectively.

One notable example of this debate is the "specious present" concept proposed by William James. James argued that the present moment is not a mathematical instant but rather an experiential moment that extends over a brief period. This subjective experience of time, according to James, is what gives us the sense of the flow of time.

There have been numerous philosophical arguments about the nature of time, such as whether time is continuous or discrete, whether the past and future are real or simply illusions, and whether time travel is possible. These debates continue to captivate philosophers and scientists alike, as the nature of time remains one of the most intriguing and elusive concepts to understand.

The relationship between time and change has been a subject of philosophical debate for centuries. According to one school of thought, time is nothing more than the measurement of change. From this perspective, time does not exist in and of itself, but is rather a concept that we use to describe the way in which things change.

Others argue that time is a fundamental aspect of the universe, and that it exists independently of any changes that may occur within it. This view is often associated with the concept of "eternalism," which holds that all moments in time exist simultaneously, and that the passage of time is an illusion.

Another philosophical debate regarding the relationship between time and change concerns the nature of causality. Some philosophers argue that time is necessary for causality to exist, as events cannot be said to cause other events unless they occur in a certain temporal order.

Others maintain that causality can exist even in the absence of time, as long as there is a necessary relationship between the cause and effect. For example, it might be argued that the laws of physics themselves determine the relationship between cause and effect, and that time is simply a convenient way of describing this relationship.

C. The possibility of time travel

The possibility of time travel is a popular topic in science fiction, but is it really possible? In this section, we will explore the theories and possibilities of time travel.

The concept of time travel has been around for centuries, with stories of people going back or forward in time to change history, see the future, or experience events that they missed. While these stories may seem far-fetched, they have captured the imagination of people all over the world.

One of the main theories behind time travel is the idea of time dilation, which is a consequence of Einstein's theory of relativity. According to this theory, time is not absolute but relative to the observer. This means that time can appear to pass differently for two people in different locations, depending on their speed or proximity to a massive object such as a black hole.

This phenomenon has been observed and measured using atomic clocks on airplanes and satellites, which have shown that time does indeed pass more slowly at higher speeds or in stronger gravitational fields. While this time dilation is small in everyday life, it could become significant for objects moving at speeds close to the speed of light or in the vicinity of a black hole.

Another theory of time travel involves the concept of wormholes, which are hypothetical shortcuts through space-time that could allow a person to travel from one point in time to another. Wormholes are predicted by Einstein's theory of relativity, but they have not been observed or proven to exist.

However, even if wormholes exist, they would be extremely unstable and would require enormous amounts of energy to keep them open long enough for a person to travel through them. The energy required is so vast that it is currently beyond our technological capabilities, making time travel via wormholes an unlikely possibility at present. There are also many paradoxes and complications that arise when considering time travel. The most famous of these is the grandfather paradox, which asks what would happen if a time traveller were to go back in time and kill their own grandfather, preventing their own birth. This paradox and others like it have yet to be resolved and remain open questions in the field of theoretical physics.

Despite the many obstacles and unknowns, the concept of time travel continues to captivate the public imagination and inspire scientific inquiry. While we may never know for sure whether time travel is possible, the theories and debates surrounding it are sure to continue for generations to come.

D. THE NATURE OF THE *present moment*

The concept of the present moment is a subject of philosophical debate, with different views on its nature and existence. In this chapter, we will explore the different perspectives on the nature of the present moment.

One view is the "moving spotlight" theory, which suggests that the present moment is constantly moving forward in time and that there is a unique "now" that moves along with it

The moving spotlight theory is an interesting concept in the philosophical debate about the nature of the present moment. According to this theory, the present moment is not a fixed point in time, but rather a moving spotlight that continuously moves along the timeline.

In other words, the present moment is constantly shifting along the timeline, illuminating different events at different times. This means that what is considered to be the present moment is always changing and is never fixed.

The moving spotlight theory challenges the idea that the present moment is a fixed and objective reality. Instead, it suggests that the present moment is a subjective experience that is dependent on the observer's perspective and context.

One of the implications of the moving spotlight theory is that the past and the future do not exist in the same sense that the present moment exists. The past and the future are only defined in relation to the present moment, which is constantly moving and changing.

The moving spotlight theory has been debated among philosophers for decades. Some argue that it provides a better account of the nature of time and the present moment than other theories, while others reject it as being too radical and implausible.

Despite the ongoing debate, the moving spotlight theory remains an intriguing concept that challenges our traditional understanding of time and the nature of reality.

Another view is the "growing block" theory, which suggests that the past and present exist but the future does not. According to this view, the present moment is constantly growing as more events enter into the past, creating a block of events that exist. This theory suggests that time is like a growing block, where the present is the leading edge of the block and the past is everything that has already happened, and the future is the part of the block that has yet to be added.

According to the growing block theory, as time passes, the block grows larger and the past becomes more extensive. The present moment, then, is not an isolated point in time but rather a slice of the growing block that includes all that has happened up to that point.

One of the main challenges to the growing block theory is the problem of how to account for the experience of the passage of time. If the past and present exist but the future does not, then how can we explain our sense that time is constantly moving forward? Some philosophers have suggested that the growing block theory can be reconciled with our experience of time by suggesting that the present moment is the constantly changing boundary between the past and the future.

A third view is the "eternalism" theory, which suggests that all points in time, including the past, present, and future, exist in a timeless block. According to this view, the present moment is no more real than any other point in time, and there is no unique "now" that moves along in time. Instead, all moments in time exist simultaneously, and our experience of time is simply an illusion

According to the eternalism theory, all moments in time are equally real, just as all points in space are equally real. This means that the past, present, and future are all equally real and exist independently of our perception of them.

Eternalism is often associated with the concept of "block time," in which time is viewed as a static four-dimensional block that contains all events, past, present, and future. From this perspective, there is no objective present moment or "now." Instead, all moments in time exist simultaneously and independently of one another, just as all points in space exist independently of one another.

One of the main challenges to eternalism is the issue of change. If all moments in time exist simultaneously, then how can change occur? Eternalists argue that change is a matter of perspective, and that what we perceive as change is simply a shift in our perspective within the four-dimensional block of time. For example, a person's birth, life, and death all exist simultaneously within the block, but our perception of time gives us the illusion of progression and change.

The eternalism theory has implications for many areas of philosophy, including ethics, metaphysics, and the philosophy of science. It challenges our intuitions about the nature of time and raises questions about the reality of the past, present, and future, and our place within the fabric of time.

The connection between time and consciousness

The relationship between time and consciousness is a fascinating and complex topic that has captured the attention of philosophers, neuroscientists, and physicists for centuries. Time is a fundamental aspect of our experience, and yet our understanding of its nature is far from complete. In this chapter, we will explore some of the key debates and ideas about the relationship between time and consciousness.

One of the key questions in this area is whether time is an objective feature of the world, or whether it is a product of our subjective experience. Some philosophers argue that time is a purely subjective experience that is created by our brains, while others contend that time is a real and objective feature of the universe that exists independently of our perception of it.

One of the most famous arguments for the subjective nature of time was put forward by the philosopher Immanuel Kant. Kant argued that time is a necessary precondition for our perception of the world. In other words, we cannot experience anything without first experiencing time. However, he also believed that time is not an objective feature of the world, but rather a subjective framework that we impose on our experiences.

Another philosopher, Henri Bergson, argued that time is a fundamentally subjective experience. Bergson believed that time cannot be understood using scientific or mathematical methods, but rather must be approached through intuition and subjective experience. According to Bergson, our experience of time is always changing, and our perception of the present moment is shaped by our past experiences.

In contrast to these subjective views of time, there are also philosophers who argue that time is a real and objective feature of the world. One of the most prominent proponents of this view was the philosopher Gottfried Wilhelm Leibniz. Leibniz believed that time is an intrinsic feature of the universe, and that it is an essential part of the fabric of reality.

In recent years, neuroscientists have made significant progress in understanding the relationship between time and consciousness. Research has shown that our perception of time is influenced by a range of factors, including our mood, attention, and memory. For example, studies have shown that time seems to pass more quickly when we are having fun, and more slowly when we are bored or anxious.

Some researchers have also suggested that our experience of time is linked to the way in which our brains process information. The brain appears to create a continuous stream of conscious experience by rapidly integrating and processing sensory information from different sources. This process creates the illusion of a seamless and continuous present moment.

One of the most intriguing and controversial ideas about the relationship between time and consciousness is the concept of "time dilation." According to this idea, time can appear to pass more slowly or quickly depending on our state of consciousness. This phenomenon has been observed in a range of contexts, from extreme sports to meditation and psychedelic experiences.

The concept of time dilation is closely related to the idea of altered states of consciousness. Altered states of consciousness can be induced by a range of techniques, including meditation, hypnosis, and psychedelic drugs. Some researchers have suggested that these altered states of consciousness can provide insights into the nature of time and consciousness.

The subjective experience of time

Despite the fact that time seems to be an objective and measurable phenomenon, our experience of it can vary greatly depending on the situation, our mental state, and even our cultural background.

One of the most interesting aspects of the subjective experience of time is the fact that it can seem to speed up or slow down depending on our circumstances. For example, when we are engaged in an enjoyable activity, such as watching a movie or spending time with friends, time can seem to fly by. Conversely, when we are in a boring or unpleasant situation, time can seem to drag on endlessly.

There are many theories as to why this might be the case. Some scientists believe that our perception of time is influenced by the amount of information we are processing at any given moment. When we are engaged in an activity that requires a lot of mental processing, such as learning a new skill or playing a complex game, time can seem to slow down because our brain is working harder and taking in more information. On the other hand, when we are engaged in a more passive activity, such as watching TV, time can seem to speed up because our brain is not being challenged to the same extent.

Another factor that can influence our perception of time is our emotional state. When we are feeling stressed or anxious, time can seem to move more slowly because we are hyper-aware of the passing of time and the pressures we are under. Similarly, when we are feeling happy and relaxed, time can seem to move more quickly because we are less focused on the passing of time and more focused on enjoying the moment.

One popular explanation for the subjective experience of time is that it is related to the way in which the brain processes information. According to this theory, the brain processes information in discrete chunks or "time slices" that are then stitched together to create a sense of continuous time. This stitching process is thought to be responsible for the subjective experience of time passing.

Another theory suggests that the subjective experience of time is related to attention and the way in which we focus on different aspects of our environment. When we are engaged in a stimulating activity or focused on a particular task, time appears to pass more quickly. Conversely, when we are bored or waiting for something to happen, time seems to drag on.

There are also cultural and societal factors that can influence the subjective experience of time. For example, some cultures have a more relaxed attitude towards time, while others place a greater emphasis on punctuality and efficiency. In addition, societal changes such as the acceleration of technology and the pace of modern life can have an impact on our subjective experience of time.

The subjective experience of time is also closely linked to our emotional states. Studies have shown that when we are experiencing positive emotions, time appears to pass more quickly. Conversely, when we are experiencing negative emotions such as pain or boredom, time seems to drag on.

While there is still much to be learned about this fascinating subject, research in this area has the potential to provide new insights into the nature of consciousness and the human experience.

The role of memory

It is often said that memory is what gives us a sense of continuity over time, allowing us to connect our past experiences to our present and future selves. Memory allows us to keep track of past events and experiences, and to use them to make predictions about the future. Without memory, we would have no sense of the passage of time, and our lives would seem like an endless stream of disconnected moments.

One influential theory about the relationship between memory and time is the "prospective memory" theory, which suggests that our ability to remember to do things in the future is central to our experience of time. According to this theory, we use past experiences to form "memory cues" that help us remember to perform a particular action at a particular time. For example, we might remember to take our medication in the morning because we associate it with the act of brushing our teeth.

One way in which memory influences our experience of time is through the phenomenon of "mental time travel." This refers to our ability to mentally project ourselves into the past or future, through memory or imagination. When we remember a past event, for example, we are able to mentally transport ourselves back in time and re-experience aspects of that event in our minds. Similarly, when we imagine a future event, we are able to mentally project ourselves into the future and anticipate what it might be like.

The relationship between memory and time is not always straightforward. For example, the "time paradox" refers to the phenomenon where time seems to speed up as we get older. This can be explained in part by the fact that our brains tend to process familiar information more quickly, which means that as we accumulate more experiences, each individual experience seems to take up less "mental time."

Another way in which memory influences our experience of time is through the way in which we perceive the passage of time. Studies have shown that our perception of time can be distorted by the amount of information we are able to remember about a particular period of time. For example, if we remember more details about a particular day, we may perceive it as having lasted longer than a day about which we remember very little.

Memory also plays a key role in shaping our sense of identity over time. Our memories of past experiences help to shape who we are in the present, and our anticipation of future experiences helps to shape who we will become in the future. Without memory, our sense of self would be constantly shifting and unstable.

THE RELATIONSHIP BETWEEN time and free will

The relationship between time and free will is a longstanding philosophical debate. At the heart of this debate is the question of whether time is predetermined or whether it allows for free will.

Determinism, the idea that every event is predetermined and caused by previous events, suggests that free will is an illusion. According to determinism, everything that happens in the future is already set in stone, and we are simply living out the consequences of actions that have already taken place. This view of time and the universe suggests that we have no control over our lives, and that everything we do is predetermined by past events.

In contrast, free will suggests that we have the ability to make choices and act independently of external factors. This view of time and the universe suggests that the future is not predetermined and that we have the ability to shape it through our choices and actions.

The debate between determinism and free will has significant implications for our understanding of time. If determinism is true, then time is simply a linear sequence of events that has already been set in motion. The future is predetermined and cannot be changed. If free will is true, then time is a much more open and flexible concept. The future is not set in stone and can be shaped by our choices and actions.

One way to understand the relationship between time and free will is through the concept of branching time. According to this view, every time we make a choice, the universe splits into multiple branches, each representing a different possible outcome. This suggests that the future is not predetermined, but rather a vast array of possible outcomes that are constantly being created by our choices and actions.

Another way to understand the relationship between time and free will is through the concept of "presentism." Presentism is the view that only the present moment exists, and that the past and future are simply ideas that we create in our minds. This view suggests that time is not predetermined, but rather a constantly evolving present moment that we have the ability to shape through our choices and actions.

The impact of time on human life and experience

Time is an inseparable part of human life and experience. It shapes our daily routines, governs the seasons and tides, and provides a framework for our personal and collective histories. The impact of time on human life is multifaceted, ranging from the physical changes that occur as we age to the social and cultural shifts that shape our worldviews and aspirations. This chapter explores some of the ways in which time influences our lives and experiences, both on an individual and societal level.

Time and Aging

One of the most profound ways in which time affects human life is through the process of aging. As we grow older, our bodies and minds undergo a series of changes that can affect our health, mobility, and cognitive abilities. Age-related changes in the brain can lead to declines in memory, attention, and executive function, while physical changes can affect balance, vision, and hearing. These changes can have a significant impact on our daily lives, affecting our ability to work, socialize, and enjoy leisure activities.

Time and Culture

Time is not just a physical phenomenon, but a cultural one as well. Different cultures have distinct attitudes and beliefs about time, which can affect how individuals perceive and experience it. For example, some cultures prioritize punctuality and schedules, while others place a greater emphasis on social relationships and flexibility. These cultural differences can lead to misunderstandings and conflicts when people from different backgrounds interact. They can also shape broader societal norms and expectations, such as the age at which people are expected to marry, have children, or retire.

Time and Memory

Memory is intimately connected to the experience of time. Our memories help us to create a sense of continuity and coherence across different moments in our lives, allowing us to form a narrative of our personal histories. However, memories are not always accurate or reliable, and the passage of time can affect how we remember events. Research suggests that memories can become distorted or forgotten over time, and that our recollections of past events can be influenced by our current emotional and cognitive states.

TIME AND TECHNOLOGY

The rise of digital technology has fundamentally changed the way that we experience time. The internet, smartphones, and social media have made it possible to connect with people and information instantaneously, breaking down the barriers of geography and time zones. However, these technologies have also created new challenges, such as the blurring of boundaries between work and leisure time, the pressure to be constantly available and responsive, and the potential for information overload and burnout.

Time and Death

The ultimate impact of time on human life is mortality. The fact that our time on earth is limited is a source of both fear and inspiration for many people. The awareness of our own mortality can prompt us to live more fully in the present moment, to seek out meaningful experiences, and to strive for personal growth and self-actualization. However, it can also lead to existential angst and anxiety, as we confront the inevitability of our own demise.

THE PHYSICS OF TIME
The nature of time in physics

In the realm of physics, time is considered one of the fundamental quantities along with mass and length. The concept of time has undergone significant changes from its initial understanding as a linear, absolute entity. The nature of time in physics is described by various theories, each of which presents a unique perspective on the nature of time.

One of the most significant contributions to our understanding of time in physics came from Albert Einstein, who proposed his theory of relativity in the early 20th century. According to Einstein's theory, time is not a universal constant, but rather a relative quantity that depends on the observer's reference frame. This theory challenged the classical view of time as an absolute and fixed quantity.

The theory of relativity introduced the concept of spacetime, which is a four-dimensional space-time continuum that combines space and time into a single entity. This concept is central to understanding the nature of time in physics. In this view, time is a dimension that is intertwined with space, and it is impossible to separate the two.

Another concept that is important in understanding the nature of time in physics is the concept of entropy. Entropy is a measure of the disorder or randomness of a system, and it is closely linked to the concept of time. The second law of thermodynamics states that the entropy of a closed system will tend to increase over time. This means that as time passes, systems tend to become more disordered, and this has a significant impact on the nature of time.

The concept of time dilation is also crucial in understanding the nature of time in physics. Time dilation refers to the effect that gravity has on time. According to Einstein's theory of general relativity, gravity warps space-time, and this can cause time to pass at different rates in different regions of space. For example, time passes more slowly near a massive object such as a black hole.

In quantum mechanics, time plays a critical role in the behaviour of particles. According to the Copenhagen interpretation of quantum mechanics, the state of a particle is determined by a wave function that evolves over time. The nature of time in quantum mechanics is a topic of ongoing research and debate, and several theories have been proposed to explain it.

Time in Classical Physics:

Time in classical physics refers to a concept that has been used for centuries to describe the behaviour of objects in the physical world. Classical physics is based on the idea that time is a constant and universal quantity that flows uniformly and independently of the observer. This concept of time has been essential to the development of classical mechanics, which deals with the behaviour of objects in motion.

In classical physics, time is typically considered to be an absolute quantity that exists independently of any physical processes. This means that time is seen as an independent parameter that can be measured and compared across different events. For example, the duration of a motion or the time it takes for an object to travel a certain distance can be measured using standard units of time, such as seconds or minutes.

One of the key features of classical physics is the idea that time is reversible, which means that the laws of physics work equally well when time runs forwards or backwards. In other words, the behaviour of objects in motion is symmetric with respect to time. This symmetry is evident in the laws of mechanics, such as Newton's laws of motion, which describe the behaviour of objects in motion in terms of the forces acting on them.

Another important aspect of classical physics is the idea of locality, which refers to the fact that physical interactions between objects are limited to their immediate surroundings. In other words, the behaviour of objects at one point in space is independent of the behaviour of objects at a distant point in space. This principle is fundamental to classical mechanics, and it underlies many of the most important ideas in classical physics.

TIME IN MODERN PHYSICS:

In contrast to classical physics, modern physics has a different view of time. In the early 20th century, Einstein revolutionized our understanding of time with his theory of relativity. According to Einstein's theory, time is relative and depends on the observer's motion and the strength of the gravitational field they experience.

In the special theory of relativity, Einstein showed that time and space are not absolute, but rather depend on the relative motion of the observer and the observed. Specifically, he showed that the faster an object moves, the slower time appears to pass for it compared to a stationary observer. This effect is known as time dilation and has been confirmed by numerous experiments, including the famous Hafele-Keating experiment, which used atomic clocks on airplanes to measure time dilation.

The Hafele-Keating experiment is a famous experiment conducted in 1971 by Joseph C. Hafele and Richard E. Keating, two physicists from the United States. The experiment was designed to test the effect of time dilation predicted by Einstein's theory of relativity.

The experiment involved four atomic clocks, which were flown on commercial airliners around the world. The planes travelled in opposite directions and covered a distance of approximately 18,000 miles each. The experimenters compared the times recorded by the clocks with the times recorded by a master clock on the ground. The master clock was used as a reference to calculate the time dilation predicted by relativity.

The experiment showed that the clocks on the planes had experienced time dilation, as predicted by relativity. The clocks had measured a time that was slightly different from the master clock on the ground. The time difference was consistent with the predicted effect of time dilation due to the speed and altitude of the planes.

The Hafele-Keating experiment provided a clear demonstration of the validity of Einstein's theory of relativity, which had been the subject of much debate and controversy in the scientific community. The experiment showed that time is not an absolute quantity but is relative to the observer's motion and the gravitational field in which the observer is located.

The experiment also had practical implications for technologies such as the Global Positioning System (GPS), which relies on accurate time measurements to determine location. The results of the experiment showed that time dilation due to the speed and altitude of GPS satellites must be taken into account for accurate location calculations.

Einstein's general theory of relativity extended these ideas to include gravity. In general relativity, the curvature of spacetime is related to the distribution of matter and energy. This curvature affects the passage of time, so time appears to pass more slowly in regions of strong gravitational fields. This effect is known as gravitational time dilation and has been observed in numerous experiments, such as with the orbiting clocks of the GPS system.

Quantum mechanics, the other major pillar of modern physics, also has a unique view of time. In quantum mechanics, time is treated as a parameter that is separate from space, unlike in relativity where time and space are combined into spacetime. The Schrödinger equation, which governs the behaviour of quantum particles, is deterministic and describes how the wave function of a system evolves over time. However, the measurement process in quantum mechanics is probabilistic, and the collapse of the wave function appears to be instantaneous,raising questions about the nature of time in quantum mechanics.

Measuring Time:

Measuring time is an essential part of our daily lives, from waking up in the morning to catching a train, attending a meeting, or even cooking a meal. Throughout history, various methods have been used to measure time, from ancient sundials and water clocks to modern atomic clocks. In modern times, we rely on highly accurate and precise time measurements in fields such as navigation, telecommunications, and scientific research.

The most basic unit of time measurement is the second, which is defined as the duration of 9,192,631,770 periods of the radiation corresponding to the transition between the two hyperfine levels of the ground state of the cesium-133 atom. This definition of the second is used to define the International System of Units (SI), which is the standard system of measurement used in scientific research and international commerce.

There are various devices and methods used to measure time, and they can be broadly classified into two categories: clocks and chronometers. A clock is a device that measures time continuously, while a chronometer is a clock that has been certified to meet certain accuracy standards. Some common examples of clocks include wall clocks, alarm clocks, and wristwatches, while chronometers include marine chronometers and atomic clocks.

The earliest devices used to measure time were sundials and water clocks. Sundials use the position of the sun to mark the hours, and they were commonly used in ancient civilizations such as Egypt and Greece. Water clocks, on the other hand, measure time by the flow of water from a container with a small hole to another container with markings indicating the hours. They were used in ancient China, India, and Greece, and were popular until the development of mechanical clocks in the Middle Ages.

Mechanical clocks were first developed in Europe in the 13th century, and they used a pendulum or balance wheel to regulate the timekeeping. The invention of the quartz clock in the 1920s revolutionized timekeeping, and quartz clocks and wristwatches are now widely used for personal timekeeping.

In modern times, atomic clocks are the most accurate timekeeping devices, with accuracy to within a few billionths of a second per day. They use the oscillations of atoms, typically caesium or rubidium, to measure time, and they are used in scientific research and in applications such as GPS navigation.

Spacetime:

Spacetime is a fundamental concept in modern physics that combines the traditional notions of space and time into a single four-dimensional continuum. It was first introduced by Hermann Minkowski in 1908, who proposed that space and time were inseparable and that the geometry of the universe could be described in terms of a four-dimensional spacetime fabric. This concept was further developed by Albert Einstein in his theory of general relativity, which posits that spacetime is not a static, unchanging background but is itself affected by the presence of massive objects.

In the framework of general relativity, spacetime is described as a curved manifold that is influenced by the distribution of matter and energy. Objects with mass cause spacetime to curve around them, altering the path of objects that travel through that region. This curvature of spacetime is what we experience as gravity.

Spacetime can be represented mathematically using a coordinate system, similar to the way we represent points in space using a Cartesian coordinate system. The coordinates of an event in spacetime consist of three spatial coordinates (x, y, z) and a time coordinate (t), which specify the location and time of the event relative to a chosen reference frame.

The concept of spacetime has important implications for our understanding of the universe. For example, it helps to explain the phenomenon of time dilation, where time appears to move slower in regions of strong gravitational fields. It also plays a central role in the development of theories such as string theory and loop quantum gravity, which seek to reconcile general relativity with quantum mechanics.

The connection between time, space, and matter

In modern physics, time is considered to be intimately connected with space and matter, forming a fundamental fabric of the universe known as spacetime. According to the theory of relativity, space and time are not separate and independent entities, rather they are intertwined in a single four-dimensional fabric.

This means that any physical event can be described in terms of its location in both space and time, as well as the properties of the matter or energy involved. In other words, spacetime provides a framework for understanding the behaviour of the universe at both the smallest and largest scales.

One of the key consequences of this view is that time can be affected by both gravity and motion. For example, time appears to slow down in the presence of a strong gravitational field, such as that near a massive object like a black hole. Similarly, time dilation occurs when an object moves at a high velocity relative to an observer, causing time to appear to pass more slowly for that object.

These effects are not just theoretical curiosities but have been confirmed through a range of experiments and observations. The GPS system, for example, relies on precise measurements of time that are corrected for relativistic effects due to the satellites' motion and position in Earth's gravitational field.

The concept of spacetime arises from the idea that the laws of physics are invariant under transformations of space and time. This invariance is expressed mathematically in terms of the spacetime interval, which is the distance between two events in spacetime. In special relativity, the spacetime interval is given by:

$ds^2 = -c^2dt^2 + dx^2 + dy^2 + dz^2$,where c is the speed of light, it is time, and x, y, and z are the spatial coordinates. The minus sign in front of the time term reflects the fact that time and space have opposite signs in the spacetime interval.

The concept of spacetime has important implications for our understanding of the behaviour of matter and energy. In special relativity, the motion of an object through spacetime is described by its four-velocity, which is a vector that describes the object's motion through both space and time. The energy and momentum of the object are also described by four-vectors that are intimately connected with its four-velocity.

In general relativity, the connection between spacetime and matter is even more profound. In this theory, matter and energy warp the fabric of spacetime, creating what we perceive as gravitational forces. The curvature of spacetime is described by the Einstein field equations, which relate the curvature of spacetime to the distribution of matter and energy within it.

The idea of a unified spacetime has also had profound implications for our understanding of the nature of time itself. In special relativity, the notion of a universal "now" is undermined by the fact that the order of events depends on the observer's frame of reference. In general relativity, the curvature of spacetime can even create closed time like curves, which would allow for the possibility of time travel.

The implications of quantum mechanics for our understanding of time

The implications of quantum mechanics for our understanding of time are still an area of active research and debate. One of the most notable implications of quantum mechanics is the phenomenon of quantum entanglement, which seems to suggest that particles can be instantaneously connected, regardless of the distance between them. This has led to the proposal of non-local interpretations of quantum mechanics, which could have implications for our understanding of time.

In classical physics, an object can only exist in one place at a time. However, in quantum mechanics, objects can exist in multiple places at once, a phenomenon known as superposition. This means that particles don't have a definite position until they are observed, and the act of observation can affect the behaviour of particles.

This has led some physicists to suggest that time may not be as fundamental as we once thought. Instead, time may be an emergent property that arises from the interactions of particles in the universe. This idea is called the "block universe" or "static universe" view of time.

According to this view, the past, present, and future all exist simultaneously, much like a static block of space-time. This means that time travel may be possible, as one could theoretically travel to any point in space-time.

However, this idea is still highly debated and controversial within the physics community. Other physicists argue that time is a fundamental aspect of the universe and cannot be reduced to the interactions of particles alone.

Another implication of quantum mechanics is the concept of "quantum entanglement." When two particles are entangled, their states are linked, and the measurement of one particle affects the state of the other, no matter how far apart they are. This phenomenon has been observed experimentally, and has led some scientists to suggest that time might be an emergent property of quantum entanglement.

The basic idea is that entanglement, which is a quantum phenomenon whereby two particles can become connected in such a way that their properties become correlated, might be the key to understanding the nature of time. The argument goes that if entanglement can be used to create a kind of "quantum clock" that ticks at a specific rate, then this could provide a way to measure time.

This theory is based on the idea that the fabric of spacetime is made up of tiny bits of information, which are encoded in the quantum states of particles. When these particles become entangled, they effectively share their information with each other, and this creates a kind of "quantum clock" that ticks at a specific rate.

According to this theory, the apparent flow of time is actually an illusion that arises from the way in which our brains process information. In other words, time is not a fundamental aspect of the universe, but rather an emergent property that arises from the complex interactions of particles.

While this theory is still in its early stages, it has the potential to revolutionize our understanding of time and the universe as a whole. However, there is still much research that needs to be done in order to fully explore the implications of this idea and to determine whether it is indeed a viable explanation for the nature of time.

TIME AS A DIMENSION
The concept of dimensions in physics

Time as a dimension is a fundamental concept in physics and plays a crucial role in our understanding of the universe. In traditional Euclidean geometry, there are three dimensions: length, width, and height. However, in physics, time is also considered as the fourth dimension, and this concept is central to the theory of relativity and other areas of physics.

The idea of time as a dimension is based on the concept of spacetime, which is a four-dimensional space where each event is described by its position in space and time. According to this view, an event can be described by four coordinates: three spatial coordinates and one temporal coordinate. This means that time is not considered as something separate from space but rather an integral part of it.

The concept of time as a dimension is closely related to the concept of time dilation, which is a phenomenon predicted by Einstein's theory of relativity. According to this theory, time dilation occurs when an object moves at a significant fraction of the speed of light. This means that time passes slower for objects in motion than for stationary objects. The faster an object moves; the slower time passes for it.

Another important consequence of considering time as a dimension is the concept of time travel. In principle, if time is considered as a dimension, it is possible to move forward or backward in time, just as we can move in any other direction in space. However, this idea remains largely theoretical, and there is currently no known way to travel through time.

The concept of time as a dimension also has implications for our understanding of the universe's structure and evolution. According to the Big Bang theory, the universe began as a singularity, a point of infinite density and temperature. As the universe expanded, it created space and time. Therefore, time is not something that exists independently of the universe but is an integral part of its structure.

In cosmology, time is considered as the fourth dimension, along with the three spatial dimensions. This concept of time as a dimension has profound implications for our understanding of the universe and its evolution.

According to the prevailing model of the universe, the Big Bang, time and space were created together in a single event, and the universe has been expanding ever since. This means that as the universe expands, the distance between objects increases, and so does the time it takes for light or any other signal to travel between them. This effect is known as cosmological redshift, and it is a crucial piece of evidence in support of the Big Bang theory.

One of the consequences of the expansion of the universe is that it is not possible to observe all of it. There are parts of the universe that are so far away that their light has not had time to reach us yet. This means that our observations of the universe are limited to a region called the observable universe, which is estimated to be about 93 billion light-years in diameter.

The concept of time as a dimension also plays a key role in the theory of relativity. According to this theory, the flow of time is not absolute but depends on the observer's reference frame. This means that time can appear to pass more slowly for a moving observer than for a stationary observer. This effect is known as time dilation, and it has been observed in experiments with high-speed particles.

It allows us to describe the expansion of the universe, the limitations of our observations, and the effects of motion on the passage of time.

The arguments for and against time being considered a dimension

The concept of time as a dimension first arose in the early 20th century with the development of Einstein's theory of relativity. Einstein's theory showed that space and time were not separate entities but were intimately connected in a four-dimensional space-time continuum. In this view, time is just another dimension of space, albeit one that behaves differently from the other three dimensions. This idea of space-time as a single, unified entity has become a cornerstone of modern physics and cosmology.

One argument for time being considered a dimension is based on the fact that we experience time as a dimension-like phenomenon. We move through time in a continuous and uniform way, just as we move through space. Time seems to have the same kind of structure as the three dimensions of space, with past, present, and future analogous to up-down, left-right, and forward-backward. This suggests that time is not just a concept or a measure of duration but a real dimension of the universe.

Another argument in favour of time being a dimension is based on the equations of physics. In most physical theories, time appears as a coordinate on an equal footing with the three dimensions of space. The equations of motion and other physical laws are formulated in terms of space-time coordinates, which treat time as a dimension alongside the three dimensions of space. This suggests that time is not just a mathematical convenience but a fundamental aspect of the universe.

However, there are also arguments against time being considered a dimension. One of the main arguments is that time behaves differently from the three dimensions of space. Unlike space, time only moves in one direction, from past to future. This directionality of time is a crucial feature of our experience of the world, and it is not clear how it can be accommodated within the concept of time as a dimension. If time is a dimension like the others, why can't we move freely in both directions along the time axis, just as we can move freely in any direction in space?

Another argument against time being a dimension is that it is not clear what it means for time to be a dimension. The three dimensions of space have a clear physical interpretation: they are the directions in which we can move and measure distances. But what is the physical interpretation of the time dimension? What does it mean to move in time, and what is the physical significance of the distance between two points in time? These questions are difficult to answer, and it is not clear whether they have any meaningful answers at all.

One of the primary arguments against time being considered a dimension is the fact that time is fundamentally different from space. While space is continuous and infinite, time is discrete and measured in distinct units, such as seconds or years. This discreteness makes it difficult to treat time as a dimension in the same way as space, as there is no continuous "flow" of time in the same way that there is a continuous flow of space.

Additionally, the directionality of time sets it apart from space as well. While we can move forwards and backwards in space, we are only able to experience time moving in one direction: forward. This asymmetry of time, known as the "arrow of time," is a fundamental aspect of our experience of the world, and it makes it difficult to treat time as a dimension that is symmetrical with space.

However, there are several arguments in favour of treating time as a dimension. One of the most compelling is the fact that time is a fundamental component of our most successful theories of the universe, including Einstein's theory of relativity and quantum mechanics. In these theories, time is treated as a dimension alongside space, and the equations governing the behaviour of matter and energy depend on this treatment.

Additionally, the concept of spacetime, which combines time and space into a single entity, has been shown to be a useful tool for understanding the behaviour of the universe. In this view, time and space are not separate entities, but rather two aspects of a unified whole. This perspective has been essential in developing our understanding of the universe on both the largest and smallest scales.

Another argument in favour of time as a dimension is the fact that it allows for a consistent framework for describing the behaviour of objects in motion. In the absence of time as a dimension, it would be difficult to account for the fact that objects change position over time. By treating time as a dimension, we are able to describe the motion of objects in a consistent way that accounts for both their position and their velocity over time.

Despite these arguments, the debate over whether time should be considered a dimension remains unsettled. Some physicists and philosophers argue that treating time as a dimension leads to inconsistencies and paradoxes, and that we need a fundamentally new way of thinking about time in order to fully understand it. Others argue that the concept of time as a dimension is essential for our understanding of the universe, and that any attempts to do away with it would be misguided. As we continue to explore the mysteries of the cosmos, we will undoubtedly continue to grapple with these complex and fascinating questions about the nature of time and its relationship to space and matter

The potential implications of time as a dimension for our understanding of the universe

If time is indeed a dimension, then it suggests that the past, present, and future all exist simultaneously and are equally real. This view is known as eternalism or the block universe theory, which posits that the universe is a four-dimensional block of spacetime in which all events, past, present, and future, are equally real.

One of the implications of the block universe theory is that determinism may be true. Determinism is the philosophical idea that every event or action has a cause, and that given the causes, the effects will necessarily follow. If the future already exists as part of the four-dimensional block of spacetime, then it seems to follow that there is no free will, as everything is predetermined.

However, not everyone agrees with this view. Some physicists and philosophers argue that the present is special and that only the present moment is real. This view is known as presentism and posits that the future does not yet exist and the past no longer exists, leaving only the present moment as the only truly existing moment.

One of the implications of presentism is that there is no determinism, as the future does not exist yet, and therefore there is no predetermined outcome. This would suggest that free will is possible, as our choices and actions are not predetermined by the four-dimensional block of spacetime.

Another potential implication of time as a dimension is that time travel may be possible. If time is a dimension like the three spatial dimensions, then it should be possible to travel through time just as we can travel through space. However, this idea is still speculative and requires significant advances in technology and understanding of the nature of time.

The idea of time as a dimension also has implications for the study of cosmology. The expansion of the universe over time has been a key focus of cosmological research, and the idea of time as a dimension has allowed physicists to better understand how the universe has evolved over time.

Additionally, the concept of time as a dimension has led to the development of new theories, such as the theory of relativity, which have fundamentally changed our understanding of the nature of time and space.

If time is indeed considered a dimension, it would fundamentally change our understanding of the universe and the nature of reality. One of the implications of this would be that the concept of time travel would become more plausible, as we could potentially navigate through time in a similar way to how we navigate through space.

Furthermore, if time is considered a dimension, it would mean that our universe is actually a four-dimensional spacetime. This would have significant implications for our understanding of the universe, particularly in the field of cosmology.

For example, the concept of a "beginning" of the universe becomes more complex if time is considered a dimension. It would suggest that the universe didn't necessarily have a point of origin, but rather that it has always existed in some form within the four-dimensional spacetime. This is a concept that aligns with the theory of eternalism, which suggests that all moments in time exist simultaneously.

Another potential implication of time as a dimension is that it could help to unify the fundamental forces of the universe. Currently, there are four fundamental forces that govern the behaviour of matter and energy: gravity, electromagnetism, the weak force, and the strong force. However, the unification of these forces has long been a goal of physicists, as it could lead to a better understanding of the fundamental nature of the universe.

If time is considered a dimension, it could provide a framework for unifying the fundamental forces. This is because the four fundamental forces could be seen as different aspects of the curvature of spacetime, which would allow for a more coherent understanding of the universe.

The concept of time as a dimension has significant implications for our understanding of the universe, including our understanding of free will, determinism, and the possibility of time travel. It has also led to the development of new theories and has played a crucial role in the study of cosmology. While the debate over the nature of time continues, the concept of time as a dimension has revolutionized our understanding of the universe and our place in it.

TIME TRAVEL IN PHYSICS

Time travel has been a popular topic in science fiction for many years, but is it actually possible in the realm of physics? The answer is not straightforward, and depends on the specific model of physics being used.

In classical physics, time travel is generally considered impossible. This is because classical physics operates under the assumption that the laws of physics are deterministic, meaning that the future is determined by the present state of the universe. In other words, if you know the present state of the universe, you can predict the future with complete accuracy. Time travel would require the ability to change the past, which would then affect the future, creating a paradox that would violate the laws of determinism.

However, in the realm of quantum mechanics, things get a little more interesting. According to some interpretations of quantum mechanics, particles can exist in multiple states simultaneously until they are observed. This means that the future is not predetermined, and it may be possible to travel back in time and change the past without creating a paradox.

One theory that allows for time travel in quantum mechanics is the concept of a closed time-like curve (CTC). A CTC is a path through spacetime that is curved back on itself, allowing an object to return to its starting point in time. This could theoretically be achieved through the use of a wormhole, a hypothetical tunnel through spacetime that would allow an object to travel between two points in space and time instantaneously.

However, the existence of CTCs raises a number of paradoxes, such as the grandfather paradox. If you were to travel back in time and prevent your grandfather from meeting your grandmother, you would never have been born, so how could you have travelled back in time in the first place?

Another potential issue with time travel is the violation of causality. If you were to travel back in time and change something in the past, this could have unforeseeable consequences on the future. For example, killing a butterfly in the past could lead to a chain reaction of events that result in a completely different future than the one you came from.

We shall be discussing all of these in the later part of the book.

Despite the many paradoxes and issues associated with time travel, it remains an active area of research in physics. Some scientists believe that time travel may be possible in the future, but only under very specific and controlled circumstances. In the meantime, we will continue to explore the mysteries of time and its relationship to the universe.

Theoretical models of time travel in physics

The concept of time travel has captivated the human imagination for centuries, and has been the subject of countless works of fiction. However, in the realm of physics, the question of whether time travel is possible, and if so, how it might work, is a topic of serious scientific investigation.

There are currently several theoretical models of time travel in physics, each with its own set of implications. Let us try to understand them.

WORMHOLES - THEORETICAL Bridges Through Space and Time

In the popular imagination, wormholes are often depicted as tunnels or portals through which characters in science fiction stories can travel instantaneously from one point in space to another, or even to different points in time. While the idea of using wormholes for time travel may be the stuff of fiction, the concept of wormholes as a feature of the fabric of space-time is a legitimate subject of scientific investigation.

Wormholes are theoretical objects that are proposed to exist in the fabric of spacetime. They are predicted by the theory of general relativity, which describes gravity as the curvature of spacetime. A wormhole can be thought of as a shortcut through spacetime, connecting two points that are far apart in space and time. This chapter will explore the concept of wormholes, their properties, and their implications for time travel.

In the theory of general relativity, spacetime is a four-dimensional fabric that can be bent and curved by massive objects. The curvature of spacetime is responsible for the force of gravity. In the presence of massive objects, the fabric of spacetime is distorted, causing other objects to follow curved paths through spacetime. This curvature of spacetime is responsible for the phenomenon of gravitational lensing, where the path of light is bent by the gravity of massive objects.

Wormholes are predicted to exist as a consequence of the curvature of spacetime. They are theoretical objects that can be thought of as tunnels through the fabric of spacetime, connecting two distant regions of spacetime. The concept of a wormhole was first proposed by the physicists Albert Einstein and Nathan Rosen in 1935. They showed that it is possible for a wormhole to exist as a solution to the equations of general relativity.

The structure of a wormhole is such that it consists of two mouths, each located in a different region of spacetime. The mouths of a wormhole are connected by a throat, which is a narrow tunnel through spacetime. If the two mouths of a wormhole are brought close together, then a traveller passing through the throat would be able to travel from one mouth to the other in a shorter time than it would take to travel the distance between the two mouths through normal space.

The geometry of a wormhole is such that it involves a curvature of spacetime that is highly nontrivial. The throat of a wormhole must be highly curved in order to connect two distant regions of spacetime. The curvature of the throat also means that the gravitational forces acting on objects passing through the throat would be very strong. This makes it very difficult to construct a stable wormhole that could be used for time travel.

There are two main types of wormholes in physics: Schwarzschild wormholes and Morris-Thorne wormholes.

Schwarzschild wormholes are named after the German physicist Karl Schwarzschild, who first described them in 1916 as a solution to Einstein's general relativity equations. These wormholes are also known as "Einstein-Rosen bridges" because they were later studied in more detail by Einstein and his collaborator Nathan Rosen.

Schwarzschild wormholes are theoretical structures that connect two distant points in space-time, allowing for the possibility of travel between them. They are formed by the collapse of a massive star, which creates a black hole, and then a white hole, which acts as the other end of the wormhole. However, Schwarzschild wormholes are highly unstable and would require exotic matter with negative energy density to stabilize them, which has yet to be observed in nature.

The second type of wormhole, Morris-Thorne wormholes, were first proposed in 1988 by physicists Michael Morris and Kip Thorne. Unlike Schwarzschild wormholes, Morris-Thorne wormholes are not formed by the collapse of a star, but are rather constructed from exotic matter that has negative energy density. The negative energy density is required to create a "throat" or tunnel through space-time that connects two distant points.

Morris-Thorne wormholes are more stable than Schwarzschild wormholes because the negative energy density of the exotic matter stabilizes the structure. However, the exotic matter required to create a Morris-Thorne wormhole is purely theoretical and has not been observed in nature. Additionally, Morris-Thorne wormholes are highly speculative and are currently only considered as a possibility within the realm of theoretical physics.

The concept of wormholes has been the subject of much speculation and debate in the scientific community. One of the most intriguing possibilities associated with wormholes is the potential for time travel. If a wormhole could be stabilized and used for time travel, it would have profound implications for our understanding of the universe and the nature of time.

However, the idea of using wormholes for time travel is highly speculative and remains purely theoretical at this point. One of the major challenges associated with wormholes is the fact that they are inherently unstable. The strong gravitational forces acting on the throat of a wormhole mean that it would collapse almost instantly after it was created. Although there are several proposed methods for stabilizing a wormhole for time travel, but none of them have been experimentally tested or proven to be feasible. One of the main challenges in stabilizing a wormhole is preventing it from collapsing due to the extreme gravitational forces that it would generate.

One proposed method for stabilizing a wormhole is through the use of exotic matter with negative energy density, which could counteract the gravitational forces and prevent the wormhole from collapsing. However, the existence of such exotic matter is purely hypothetical and has not been observed in nature.

Another proposed method is to use the Casimir effect, which is a quantum mechanical phenomenon that results in a negative energy density in the space between two parallel plates. This negative energy density could potentially stabilize a wormhole, but it would require the plates to be very large and the energy requirements to create and maintain them would be enormous.

A third proposed method is to use a rotating black hole, which could theoretically create a stable traversable wormhole. However, the rotational energy required to stabilize the wormhole would be immense, and it is not clear if it could be generated or harnessed.

Overall, the stabilization of a wormhole for time travel remains a purely theoretical concept, and much more research is needed before it could become a practical reality.

CLOSED TIME-LIKE CURVES (CTCs)

CTCs are paths through spacetime that loop back on themselves, creating a closed loop in time. They are a theoretical concept that has been proposed by some physicists as a possible means of time travel. However, they also raise several paradoxes and inconsistencies with the laws of physics as we currently understand them. In this chapter, we will explore what closed time-like curves are, how they work, and some of the major issues surrounding them.

To understand closed time-like curves, we must first understand the concept of time-like curves. A time-like curve is a path through spacetime that connects two events, and along which a particle with mass could travel at speeds less than the speed of light. In other words, it is a trajectory that is consistent with the laws of physics as we currently understand them.

A closed time-like curve is a time-like curve that forms a closed loop in spacetime. This means that a particle could travel along the curve and end up back where it started in both space and time. If a closed time-like curve exists in a region of spacetime, then it is possible for an object or particle to travel back in time and interact with itself.

The concept of closed time-like curves was first proposed by Kurt Gödel in 1949, as a solution to the field equations of general relativity. Gödel's solution described a universe in which time travel was possible, but it raised a number of paradoxes and inconsistencies with the laws of physics.

One of the most famous paradoxes associated with closed time-like curves is the grandfather paradox. Imagine a person travels back in time and kills their own grandfather before they have any children. This would mean that the time traveller's parents would never have been born, and therefore the time traveller would not exist. This paradox and others like it have led many physicists to conclude that closed time-like curves are not physically possible.

Despite these issues, some physicists have continued to explore the concept of closed time-like curves and their potential uses for time travel. One proposed method of creating a closed time-like curve is by creating a rotating cylinder of matter, which would generate a gravitational field strong enough to warp spacetime into a closed loop. However, this would require enormous amounts of energy and exotic forms of matter that have not been observed in nature.

Another proposed method of creating closed time-like curves is through quantum mechanics. Quantum mechanics allows for the possibility of particles interacting with their own past, creating closed time-like curves. However, this has not been experimentally demonstrated and remains a theoretical possibility.

One way to avoid the paradoxes associated with closed time-like curves is to invoke the idea of parallel universes or a multiverse. In this view, a time traveller could travel back in time and interact with their past self without changing their own timeline. Instead, they would create a new timeline in a parallel universe that would diverge from the original timeline.

One possible resolution to the paradox is the Novikov self-consistency principle, proposed by physicist Igor Novikov in the 1980s. This principle suggests that closed time-like curves may be allowed in a way that ensures that any actions taken by a time traveller are consistent with the past, thus avoiding any paradoxes. According to this principle, a time traveller cannot change the past because any actions taken by the traveller have already occurred and are a part of the timeline. This means that the time traveller cannot do anything that would create a paradox.

Another proposed resolution is the many-worlds interpretation of quantum mechanics, which suggests that every possible outcome of a given event occurs in a separate parallel universe. In the case of a time traveller going back in time and potentially changing the past, this would create a new parallel universe where the time traveller's actions have altered the course of events. However, the original timeline would remain unchanged.

Despite the intriguing possibilities of closed time-like curves and the potential for time travel, the concept remains purely theoretical at this point. There is no experimental evidence to suggest that closed time-like curves or time travel are possible, and the practical implications of time travel are largely unknown. However, the study of closed time-like curves and the nature of time remains an active area of research in theoretical physics, and it is possible that new discoveries could shed light on the possibilities of time travel in the future.

While the concept remains an active area of research and speculation, it is not yet clear whether closed time-like curves are physically possible, or whether they will ever be a viable means of time travel.

COSMIC STRINGS:

Cosmic strings are hypothetical one-dimensional objects that are predicted to exist in certain models of the universe. They are similar to cosmic defects such as domain walls, monopoles, and textures, which are also predicted by certain theories. Cosmic strings are predicted to be extremely thin and can range in size from a few Planck lengths to several light-years, with a width that is believed to be smaller than the size of an atom.

The idea of cosmic strings was first introduced in the early 1970s by theoretical physicists Tom Kibble, Sheldon Glashow, and James Bjorke. They proposed that cosmic strings could be formed during phase transitions in the early universe, when the universe cooled from a state of extremely high energy to a more stable state. This idea was further developed by Alexander Vilenkin and E. P. S. Shellard in their book "Cosmic Strings and Other Topological Defects".

Cosmic strings are not to be confused with strings in string theory, which are one-dimensional objects that exist in a ten- or eleven-dimensional space-time. Cosmic strings, on the other hand, exist in the four-dimensional space-time of the universe.

One of the most intriguing aspects of cosmic strings is their potential for time travel. According to some theoretical models, cosmic strings could create a "closed time like curve" (CTC), Cosmic strings can create CTCs by bending the fabric of spacetime in such a way that the path of a particle could become a closed loop, allowing it to travel back in time. However, this idea is highly speculative and has not been proven.

One of the fascinating aspects of cosmic strings is their potential to create gravitational lensing. Cosmic strings are expected to cause distortions in the fabric of spacetime, which can cause light to bend as it travels through space. This effect can create a gravitational lensing effect, where light is bent around the cosmic string, producing multiple images of the same object.

Another interesting aspect of cosmic strings is their potential for observational evidence. Cosmic strings would have a significant effect on the distribution of matter in the universe, causing distortions in the cosmic microwave background radiation and in the distribution of galaxies. Several experiments have been designed to search for evidence of cosmic strings, but as of yet, none have been detected.

In addition to their potential for time travel and observational evidence, cosmic strings have been the subject of much theoretical research in cosmology and particle physics. They have been proposed as a possible mechanism for explaining the formation of large-scale structure in the universe, as well as for generating the seed perturbations that led to the formation of galaxies and clusters of galaxies.

Tipler cylinders:

Tipler cylinders are theoretical objects proposed by physicist Frank Tipler in 1974 that could potentially be used for time travel. These cylinders are long, infinitely dense tubes that are held together by cosmic strings, and they rotate around their longitudinal axis at speeds near the speed of light.

Tipler cylinders rely on the concept of frame dragging, which is a phenomenon predicted by Einstein's theory of general relativity. Frame dragging occurs when the rotation of a massive object causes a "dragging" effect on the space-time around it, causing other objects in its vicinity to be pulled into the rotation. In the case of Tipler cylinders, the cosmic strings would create a strong gravitational field, causing the space-time around them to warp and drag nearby objects into a closed time-like curve.

One of the main advantages of Tipler cylinders over other time travel methods is that they do not require any exotic matter or negative energy, which makes them more plausible from a theoretical perspective. However, Tipler cylinders would require an enormous amount of energy to create and stabilize, as well as the ability to manipulate cosmic strings, which are still purely theoretical.

In addition, Tipler cylinders have been criticized for violating the causality principle, which states that an effect cannot occur before its cause. The closed time-like curve created by a Tipler cylinder would allow an object to travel back in time and interact with its past self, potentially leading to paradoxes and violations of causality.

Despite these limitations, Tipler cylinders remain an intriguing concept in theoretical physics, and they continue to inspire research into the nature of space-time and the possibility of time travel. While the practical application of Tipler cylinders may be far off in the future, they represent an exciting avenue of exploration for physicists seeking to understand the fundamental laws of the universe.

The conditions under which time travel might be possible

The idea of time travel has captured the imaginations of people for centuries. The possibility of traveling through time to the past or the future has been explored in science fiction, but is it actually possible in the real world? Physicists have been exploring the possibility of time travel for decades, and have come up with several conditions under which time travel might be possible.

One of the conditions for time travel is the existence of closed time-like curves (CTCs). These are paths through spacetime that allow an object to return to its own past. The existence of CTCs is allowed in some mathematical solutions of Einstein's field equations, but they require a type of exotic matter with negative energy density that has not been observed in nature.

Another condition is the existence of wormholes, which are hypothetical tunnels through spacetime that connect two distant points in space and time. If one end of a wormhole is held stationary while the other is accelerated to near the speed of light, time dilation would occur and time travel might be possible. However, the energy requirements for creating and stabilizing a wormhole are immense, and it is not clear whether it is physically possible.

In addition to these conditions, the laws of physics would need to allow for time travel. For example, if the laws of physics do not allow for the violation of causality (the idea that cause always precedes effect), then time travel to the past might not be possible.

Finally, it is possible that time travel might require a technology or understanding of physics that we have not yet discovered. For example, some physicists have proposed the idea of "chronology protection," which suggests that the laws of physics might prevent time travel from being possible.

The potential implications of time travel for our understanding of the universe

If time travel were possible, it could have profound implications for our understanding of causality, determinism, and the possibility of altering the course of history.

One of the most significant implications of time travel is the potential to violate the principle of causality, which states that every event has a cause that precedes it in time. If time travel were possible, it could allow for events to occur without a causal explanation or for events to occur before their cause, leading to a paradoxical situation known as a causality violation. This would challenge our understanding of cause and effect and could have profound consequences for our ability to make predictions and understand the behaviour of the universe.

Another implication of time travel is the potential to change the course of history. If time travel were possible, it could allow individuals to go back in time and alter events, leading to a different future than the one that would have occurred without intervention. This raises questions about the nature of free will and the possibility of altering predetermined events. It also raises questions about the ethical implications of time travel, such as whether it would be appropriate to change the past to prevent a tragedy or whether it would be an infringement on the autonomy of those who lived through those events.

In addition to these philosophical implications, time travel could also have practical implications for our understanding of the universe. For example, it could shed light on the nature of black holes and the singularity at the centre of a black hole. It could also provide insight into the fundamental nature of time itself, including whether time is a fundamental property of the universe or an emergent property that arises from other fundamental properties.

However, it is important to note that the possibility of time travel is still purely speculative at this point. While there are theoretical models that suggest time travel may be possible under certain conditions, such as through the use of wormholes or closed time-like curves, these models are still hypothetical and have not been experimentally verified. Additionally, even if time travel were possible, there may be practical limitations that make it impossible to travel back to a specific point in time, such as the inability to create a wormhole of sufficient size or stability.

The idea of time travel raises many fascinating questions about the nature of the universe and our place in it. It challenges our understanding of causality, free will, and the fundamental nature of time itself. While there are theoretical models that suggest time travel may be possible, it remains purely speculative at this point. If time travel were to become a reality, it could have profound implications for our understanding of the universe and our place in it, but it is important to approach the subject with caution and careful consideration of its ethical and practical implications.

TIME IN ASTROPHYSICS

OBSERVING TIME IN ASTROPHYSICS

Astrophysicists measure time using various techniques and instruments, including clocks and atomic clocks, which are the most accurate timekeepers available. They also rely on observations of celestial objects, such as pulsars and supernovae, to determine the age and distance of objects in the universe.

One of the most important ways in which time is observed in astrophysics is through the study of light. The speed of light is constant, and the time it takes for light to travel from one object to another can be used to calculate the distance between them. This is known as the technique of parallax, which is used to measure the distance of nearby stars and galaxies.

Time is also observed through the study of the redshift of light from distant galaxies. As the universe expands, the wavelength of light from distant objects is stretched, which results in a redshift. By measuring the amount of redshift, astrophysicists can determine the distance and age of these objects.

One of the most basic ways to observe time in astrophysics is by measuring the period of oscillation or rotation of celestial objects. For example, astronomers can measure the rotation period of planets in our solar system by observing their surface features and tracking how they change over time. This allows them to study the planet's internal structure, as well as the influence of other objects in the solar system on its rotation.

Similarly, astronomers can measure the period of pulsation of certain stars, such as Cepheid variables and RR Lyrae stars, by tracking changes in their brightness over time. This provides information about the stars' intrinsic properties, such as their size, mass, and age, as well as their distance from Earth.

Another way to observe time in astrophysics is through the study of transient events, such as supernovae, gamma-ray bursts, and gravitational waves. These events occur over very short timescales, ranging from fractions of a second to minutes or hours, but can have profound effects on the objects and environments in which they occur.

Observing these events can provide insights into the behaviour of matter and energy in extreme conditions, as well as the structure and evolution of the universe. For example, the detection of gravitational waves from the collision of two black holes in 2015 confirmed the existence of these elusive objects, and provided new insights into the nature of gravity and the structure of gravity.

In addition to studying individual objects and events, astrophysicists also use time as a tool to study the evolution of the universe as a whole. By observing the cosmic microwave background radiation, the oldest light in the universe, astronomers can measure the age of the universe and its expansion rate over time. This information can help to constrain models of the early universe, and shed light on the nature of dark matter and dark energy, which are believed to dominate the structure and evolution of the universe at large scales.

THE ROLE OF TIME IN the Evolution of the Universe

The study of time in astrophysics is essential in understanding the evolution of the universe. The Big Bang Theory proposes that the universe began as a singularity, a point of infinite density and temperature. As the universe expanded, time began, and the universe cooled down, resulting in the formation of the first atoms and eventually galaxies.

The universe has been evolving since the Big Bang, which is estimated to have occurred around 13.8 billion years ago. Over this immense period, the universe has gone through various stages of development, including the formation of galaxies, stars, and planets.

The role of time in the evolution of the universe is a subject of great interest to astrophysicists, as it provides insights into the mechanisms that have shaped the cosmos over its vast history.

One of the most important roles of time in the evolution of the universe is in the formation of galaxies. According to the most widely accepted models, galaxies form through the gravitational collapse of clouds of gas and dust. Over time, these clouds become denser and hotter, eventually leading to the formation of stars. As stars form and begin to emit light and other forms of radiation, they create the complex structures that we observe in galaxies today. The process of galaxy formation is a slow one, occurring over billions of years, and it is heavily influenced by the properties of matter and the expansion of the universe.

Another key role of time in the evolution of the universe is in the formation and evolution of stars. Stars are formed from clouds of gas and dust that become increasingly dense over time. Once a cloud reaches a certain density, gravity takes over and begins to pull the matter together, eventually leading to the formation of a protostar. Over time, the protostar contracts and heats up, eventually becoming hot and dense enough to begin nuclear fusion. This process of fusion releases enormous amounts of energy, which causes the star to shine brightly and to create the complex chemical elements that are essential for life.

The evolution of the universe is also heavily influenced by the properties of time itself. According to the theory of general relativity, time is not absolute but is instead relative to the observer's position and motion. This means that the passage of time can vary depending on the observer's location and the strength of the gravitational field in which they are situated. In extreme cases, such as near a black hole, time can become so distorted that it appears to stand still or even to flow backwards.

The concept of cosmic time, which is the time scale on which the entire universe evolves, is also crucial to our understanding of the evolution of the universe. Cosmic time is measured by the expansion of the universe, which is currently estimated to be around 13.8 billion years old. By studying the properties of cosmic time, such as the rate of expansion of the universe, astrophysicists can gain insights into the early stages of the universe's evolution, including the formation of the first galaxies and the conditions that led to the formation of stars and planets.

One of the most significant ways in which time has played a role in the evolution of the universe is through the process of cosmic inflation. Cosmic inflation is a theoretical period of extremely rapid expansion that the universe underwent in its early stages, only fractions of a second after the Big Bang. During this time, the universe expanded by an exponential factor, increasing its size by a factor of at least 10^{26}.

This rapid expansion had profound consequences for the structure of the universe. It smoothed out any irregularities or fluctuations that existed in the early universe, making the distribution of matter and energy much more homogeneous. This is evident in the cosmic microwave background radiation, which shows that the temperature of the universe is almost exactly the same in all directions.

The process of cosmic inflation also gave rise to the formation of cosmic structures, such as galaxies and galaxy clusters. The small fluctuations that were present in the early universe were amplified by the rapid expansion, eventually leading to the formation of the large-scale structures we observe today.

Another way in which time has played a role in the evolution of the universe is through the process of nucleosynthesis. This is the process by which the light elements, such as hydrogen, helium, and lithium, were formed in the early universe. This occurred in the first few minutes after the Big Bang, when the temperature and density of the universe were extremely high.

During this time, protons and neutrons combined to form atomic nuclei, releasing large amounts of energy in the process. This energy eventually cooled down enough to allow electrons to combine with nuclei to form atoms. The resulting universe was mostly composed of hydrogen and helium, with trace amounts of other elements.

Over time, stars formed and underwent nuclear fusion, producing heavier elements such as carbon, nitrogen, and oxygen. These elements were then dispersed throughout the universe through supernova explosions and other processes. The formation of heavier elements is one of the key factors that made the evolution of life on Earth possible.

From the rapid expansion of the early universe to the formation of stars and galaxies, the passage of time has had profound consequences for the cosmos. By studying the history of the universe, we can gain a deeper understanding of the nature of time itself and our place in the cosmos.

Implications of Time in Astrophysics

The study of time in astrophysics has significant implications for our understanding of the universe. The measurement of time provides a means of determining the age and distance of objects in the universe. This, in turn, provides a means of understanding the evolution of the universe and the formation of celestial objects. The study of time in astrophysics has yielded numerous implications for our understanding of the universe. From the origin and evolution of the universe to the behaviour of stars and galaxies, time plays a fundamental role in shaping the cosmos as we know it.

One of the most significant implications of time in astrophysics is the concept of cosmic time. Cosmic time refers to the age of the universe, and the measurement of cosmic time is essential for understanding the evolution of the cosmos. The age of the universe is estimated to be around 13.8 billion years old, based on observations of the cosmic microwave background radiation and the Hubble constant. This measurement of cosmic time has led to several important discoveries, including the expansion of the universe and the existence of dark matter and dark energy.

Time also plays a crucial role in the evolution of galaxies. The formation and evolution of galaxies are driven by various physical processes that occur over vast timescales. For example, the process of star formation occurs over millions of years, and the merging of galaxies can take billions of years. The study of galaxy evolution has yielded important insights into the history of the universe, including the formation of the first galaxies and the emergence of supermassive black holes.

In addition to galaxies, time is also essential for understanding the behaviour of individual stars. The life cycle of a star is driven by various physical processes that occur over millions or billions of years. For example, the process of nuclear fusion that powers a star's energy output occurs over millions of years, and the final stages of a star's life can take billions of years. The study of stellar evolution has led to important discoveries, including the origin of elements such as carbon, oxygen, and iron.

The study of time in astrophysics has also yielded important insights into the nature of dark matter and dark energy. These mysterious substances make up most of the mass-energy content of the universe, but their properties and behaviour remain poorly understood. The study of time in astrophysics has led to various theories and models of dark matter and dark energy, including the possibility that they are related to the expansion of the universe and the behaviour of large-scale structures such as galaxies and clusters of galaxies.

Another important implication of time in astrophysics is the study of gravitational waves. These ripples in the fabric of spacetime were predicted by Albert Einstein's theory of general relativity, but it was not until recently that they were detected directly by the Laser Interferometer Gravitational-Wave Observatory (LIGO). The study of gravitational waves has provided a new way of observing the universe and has yielded important insights into the behaviour of black holes and the evolution of galaxies.

Finally, the study of time in astrophysics has important implications for the search for extra-terrestrial life. The discovery of exoplanets and the study of their atmospheres and environments is a rapidly growing field, and the search for signs of life on other planets is a fundamental question in astrobiology. The study of time in astrophysics provides important insights into the conditions necessary for the emergence and evolution of life and the potential habitability of other planets.

The study of time in astrophysics has yielded numerous implications for our understanding of the universe. From the evolution of galaxies and stars to the behaviour of dark matter and the search for extra-terrestrial life, time plays a fundamental role in shaping the cosmos. The ongoing study of time in astrophysics will undoubtedly lead to further discoveries and insights into the nature of the universe and our place within it.

The impact of gravity on the passage of time

The impact of gravity on the passage of time is one of the most fascinating phenomena in the field of physics. According to the theory of general relativity, gravity affects time in a way that is different from what we experience in our daily lives. This effect is known as gravitational time dilation.

In simple terms, the closer you are to a massive object, the slower time passes for you relative to someone further away from the object. This means that time passes more slowly in stronger gravitational fields. This effect has been observed and confirmed through various experiments, including the use of atomic clocks.

One of the most famous examples of gravitational time dilation is the experiment conducted by physicist Joseph Hafele and astronomer Richard Keating in 1971. They flew atomic clocks on commercial airplanes and found that the clocks on the planes ran slower than those on the ground due to the planes being closer to the Earth's gravitational field.

Another example of gravitational time dilation is the phenomenon of black holes. A black hole is an extremely massive object that has collapsed to a point where its gravity is so strong that nothing, not even light, can escape. As an object gets closer to a black hole, time slows down until it comes to a complete stop at the event horizon, the point of no return for anything falling into the black hole.

The impact of gravity on the passage of time also has significant implications for the evolution of the universe. According to the Big Bang theory, the universe began with a massive explosion and has been expanding ever since. The rate of this expansion is affected by the amount of matter and energy in the universe, as well as the curvature of space-time.

Gravity plays a key role in determining the curvature of space-time, and therefore the rate of expansion of the universe. If the amount of matter and energy in the universe is large enough, the gravitational pull will eventually overcome the expansion and the universe will begin to contract. This could lead to a "Big Crunch" scenario where the universe collapses back in on itself.

The study of gravitational waves, which are ripples in space-time caused by the acceleration of massive objects, has allowed scientists to study the universe in ways that were previously impossible. By observing these waves, we can learn about the behaviour of objects like black holes and neutron stars, which can help us understand the impact of gravity on the passage of time.

The impact of gravity on the passage of time is a crucial aspect of our understanding of the universe. It has been observed and confirmed through various experiments and has significant implications for the evolution of the universe. By continuing to study this phenomenon, we can gain a deeper understanding of the nature of gravity and its role in shaping the universe we live in.

The impact of gravity on the passage of time is a central concept in Einstein's theory of General Relativity. According to the theory, the force of gravity is not a force at all but rather the curvature of spacetime caused by the presence of mass or energy. This curvature affects not only the path of objects in space but also the flow of time itself.

To understand this concept, we can look at the example of a clock near a massive object like a planet or star. According to General Relativity, time is slowed down in a strong gravitational field. This means that a clock near a massive object will tick more slowly than a clock farther away from it. This phenomenon is known as gravitational time dilation.

The amount of time dilation depends on the strength of the gravitational field, which is determined by the mass and distance of the object. For example, if we were to place two identical atomic clocks side by side, one on the Earth's surface and the other in orbit around the Earth, the clock in orbit would run slightly faster due to the weaker gravitational field.

Another example of the impact of gravity on the passage of time is the phenomenon of gravitational redshift. This occurs when light is emitted from a source in a strong gravitational field and travels to a location with a weaker gravitational field. The light loses energy as it travels through the curved spacetime, resulting in a shift towards longer wavelengths (i.e., the light appears redder). This effect has been observed in the spectra of light from stars orbiting close to black holes.

The concept of gravitational time dilation has important implications for astrophysics and cosmology. For example, the fact that time runs more slowly in a strong gravitational field means that objects in the universe, such as black holes and neutron stars, can potentially exist for much longer than we might expect based on their mass and energy output.

Gravitational time dilation plays a role in the expansion of the universe. According to the theory of General Relativity, the universe is expanding and the rate of expansion is accelerating. The acceleration is believed to be caused by a mysterious force called dark energy. However, recent research has suggested that the acceleration may be due to the effects of gravitational time dilation in regions of the universe with stronger gravitational fields.

Time dilation near black holes

Time dilation is a well-known phenomenon in the realm of physics. It is a difference in the elapsed time measured by two observers due to a relative motion between them or a difference in the strength of the gravitational field between them. One of the most extreme examples of time dilation can be observed near black holes, where the gravitational pull is so strong that time appears to slow down or even stop.

Black holes are the most enigmatic objects in the universe. They are regions in space where the gravitational pull is so strong that nothing, not even light, can escape. The gravitational field near a black hole is so intense that it can warp the fabric of space-time. This curvature of space-time leads to some unusual effects, including time dilation.

Time dilation near black holes was first predicted by Albert Einstein's theory of general relativity. According to the theory, the closer an object is to a massive body, the slower time moves for it. This effect is known as gravitational time dilation. The time dilation near a black hole is so extreme that it can cause some bizarre phenomena, such as time travel and the slowing down of time.

The degree of time dilation near a black hole depends on two factors: the mass of the black hole and the distance from its event horizon. The event horizon is the point of no return beyond which the gravitational pull is so strong that not even light can escape. The closer an object gets to the event horizon, the stronger the gravitational pull becomes, and the more time slows down for it.

For example, if two clocks were placed at different distances from a black hole, the clock closer to the black hole would tick slower than the clock farther away. This effect is known as gravitational redshift. The clock closer to the black hole would appear to be running slower from the perspective of an observer farther away. This effect has been confirmed through observations of stars orbiting near black holes.

Another effect of time dilation near black holes is that it can lead to time travel. According to some theories, if a spacecraft were to approach a black hole and hover near its event horizon, time would appear to slow down so much that the spacecraft could travel thousands or even millions of years into the future in just a few years of its own time. This idea is purely theoretical, and there are many obstacles that would need to be overcome for it to become a reality.

Time dilation near black holes also has implications for the evolution of the universe. The intense gravitational fields near black holes can cause them to grow by absorbing matter from their surroundings. As more matter falls into a black hole, it becomes more massive, and its gravitational pull becomes even stronger. This can lead to the formation of supermassive black holes, which are thought to play a critical role in the evolution of galaxies.

The extreme gravitational pull near a black hole can cause time to slow down or even stop, leading to some unusual effects such as time travel and gravitational redshift. Studying the effects of time dilation near black holes can help us better understand the universe and its evolution.

The connection between time and the expanding universe

The connection between time and the expanding universe is a fundamental concept in modern cosmology. According to the current understanding of the universe, the fabric of space-time itself is expanding, and this expansion is accelerating. This has important implications for the concept of time and its relationship to the universe as a whole.

The first evidence for the expanding universe was discovered in the early 20th century by astronomer Edwin Hubble, who observed that the light from distant galaxies appeared to be shifted towards the red end of the spectrum. This redshift, he realized, was caused by the galaxies moving away from us. This discovery was the first step towards understanding the expansion of the universe.

The expansion of the universe can be described mathematically by the Hubble law, which states that the farther away a galaxy is from us, the faster it is moving away from us. This means that the universe is not only expanding, but that the expansion is accelerating.

The relationship between time and the expanding universe is complex, but it can be understood in terms of the cosmological redshift. This redshift is caused by the stretching of space-time itself, which causes the wavelength of light to increase as it travels through the expanding universe. This stretching of space-time also affects the passage of time.

In fact, the stretching of space-time near massive objects, such as galaxies and black holes, has been observed to cause time dilation. This means that time passes more slowly in regions of space-time that are affected by gravity. This effect is known as gravitational time dilation, and it has been confirmed by numerous experiments and observations.

The impact of gravitational time dilation on the expanding universe is significant. As the universe expands, the gravitational pull of massive objects becomes weaker, which means that time dilation becomes less pronounced. This has important implications for the future of the universe, as it suggests that the expansion of the universe will continue indefinitely, eventually leading to a state of maximum entropy known as the heat death of the universe.

The concept of time in the expanding universe is also closely related to the concept of the cosmological arrow of time. This arrow of time is defined by the fact that the universe appears to be evolving in a particular direction, from a state of low entropy to a state of high entropy. This is known as the second law of thermodynamics, and it implies that the universe has a definite direction in time.

The connection between time and the expanding universe is complex and multifaceted. The expansion of the universe affects the passage of time, and this effect is further complicated by the presence of gravity. Nevertheless, the observations and experiments conducted by cosmologists have provided us with a wealth of information about the nature of time in the expanding universe,

One implication of the expanding universe is that the passage of time is not uniform throughout the universe. This is known as "cosmological time dilation." Essentially, the further away an object is from us, the more it appears to be moving away from us due to the expansion of the universe. As a result, time appears to be passing more slowly in those distant regions compared to regions closer to us.

This effect has been observed through a number of different methods, such as studying the redshift of light from distant galaxies and the cosmic microwave background radiation. It has also been measured through experiments with atomic clocks on Earth and in space.

Cosmological time dilation has important implications for our understanding of the universe and its history. For example, it affects our ability to study the early universe and the formation of galaxies. By observing distant objects, we can essentially look back in time and see what the universe was like billions of years ago.

The effect of cosmological time dilation also has implications for the search for extra-terrestrial life. If life exists on a planet in a distant part of the universe, the passage of time there would be different from the passage of time on Earth. This could impact our ability to communicate with them, as well as our ability to understand the evolution of life on other planets.

In addition, the concept of cosmological time dilation is closely linked to the concept of cosmic expansion and the fate of the universe. If the expansion of the universe continues to accelerate, as current observations suggest, the effects of time dilation will become more pronounced. This has led some scientists to speculate about the ultimate fate of the universe, and whether time itself will eventually come to an end and have allowed us to deepen our understanding of the universe as a whole.

As the universe continues to expand, its expansion rate is also increasing, driven by the mysterious force of dark energy. This accelerating expansion has important implications for our understanding of time.According to the theory of relativity, the rate at which time passes depends on the curvature of spacetime. In regions of high gravitational curvature, such as near a massive object like a black hole, time passes more slowly than in regions of lower curvature. This phenomenon is known as time dilation.

In a similar way, the acceleration of the expansion of the universe can also lead to time dilation. As the universe expands, the space between galaxies is stretched, leading to a decrease in the density of matter and energy in the universe. This decrease in density results in a decrease in the gravitational curvature of spacetime, and as a result, time passes more quickly in the expanding universe than it does in a static universe.

This effect is known as cosmic time dilation, and it has been observed through the study of supernovae. Supernovae are exploding stars that can be used as standard candles to measure distances in the universe. By measuring the light curves of distant supernovae, astronomers can determine how their brightness changes over time. Because the rate at which time passes affects the observed brightness of a supernova, studying the light curves of distant supernovae can provide information about the rate of cosmic expansion and the history of the universe.

The accelerating expansion of the universe has important implications for our understanding of time. It leads to cosmic time dilation, which affects the observed brightness of distant supernovae, and it also has implications for the ultimate fate of the universe. As the expansion continues to accelerate, it is possible that the universe will eventually enter a period of exponential expansion known as the Big Rip, in which the fabric of spacetime is torn apart and everything in the universe is destroyed.

PARADOXES OF TIME

The paradoxes of time are mainly due to the human perception of time as a linear and irreversible progression. However, in the realm of physics, time is treated as a dimension that can be manipulated and its flow can be altered. This leads to various paradoxes and inconsistencies in our understanding of time.

Let us try to understand a few of them.

The grandfather paradox

The Grandfather Paradox is one of the most famous thought experiments in physics and is often discussed in the context of time travel. The paradox is named for the hypothetical scenario where a person goes back in time and kills their own grandfather before their own father was born, thereby preventing their own existence. This creates a paradox because if the person never existed, they could not have gone back in time to kill their grandfather.

The Grandfather Paradox is often used to illustrate the logical and philosophical difficulties that arise when considering the possibility of time travel. It is a prime example of what is known as a causal loop, where an event in the past causes an effect that leads back to the same event, forming a closed loop with no clear beginning or end.

There have been many proposed solutions to the Grandfather Paradox, but none of them are entirely satisfactory. One possible solution is the multiple universe theory, which suggests that when a person travels back in time and alters the course of events, they create a new parallel universe where the events of the original timeline did not occur. In this scenario, the person who travelled back in time would still exist in their own original universe, while a new parallel universe is created where their grandfather is killed and they never existed.

Another proposed solution is that the universe itself may have a self-correcting mechanism that prevents paradoxes from occurring. In this scenario, any attempt to change the past would be thwarted by some unknown force that ensures events unfold in a way that is consistent with the timeline.

Despite the proposed solutions, the Grandfather Paradox remains an unsolved problem in physics and time travel. The paradox highlights the difficulties of reconciling the idea of time travel with the laws of physics and causality, and it continues to inspire new research and discussion in the field.

The bootstrap paradox

The Bootstrap Paradox is a paradoxical situation in which an object or information appears to have no origin or source. It is also known as a causal loop or ontological paradox, and is often used in science fiction and time travel stories.

The paradox is named after the idea of "pulling oneself up by one's own bootstraps," which means to achieve something without any external help or support. In the case of the Bootstrap Paradox, this refers to the concept of creating or receiving something from oneself in the past or future without any external cause or source.

The paradox is typically presented as follows: Suppose a time traveller goes back in time and gives a famous composer a sheet of music that the composer then publishes and becomes famous for. The time traveller then goes back to the future and discovers that the sheet of music was actually written by the composer himself, who had been inspired by the time traveller's gift. In this scenario, the music has no clear origin or source, and seems to exist in a causal loop or infinite loop.

The Bootstrap Paradox raises several philosophical questions about causality, determinism, and free will. One of the main questions is whether events in the past are fixed and predetermined, or whether they can be changed by actions taken in the present or future. If events in the past are fixed and predetermined, then it seems that the time traveller could not have changed the course of history by giving the composer the sheet of music. However, if events in the past can be changed, then the paradox raises the question of what caused the time traveller to go back in time and give the sheet of music to the composer in the first place.

Another question raised by the Bootstrap Paradox is whether the existence of the sheet of music, or any other object or information, requires a specific origin or source. If an object or information can exist in a causal loop or infinite loop, does that mean that it has always existed and will always exist, without any clear beginning or end? Or does it mean that the object or information exists in a kind of timeless or eternal state, outside of linear time?

The Bootstrap Paradox has been used in numerous science fiction stories and movies, including "Doctor Who," "Terminator," and "Predestination." In many of these stories, the paradox is used as a plot device to create a sense of mystery or to explore philosophical themes related to time and causality.

The bootstrap paradox has puzzled scientists and philosophers alike because it challenges our understanding of causality and the flow of time. If an object or information can exist without any clear origin, then it raises questions about the nature of reality and whether the universe has a predetermined fate.

One proposed solution to the Bootstrap Paradox is the idea of a multiverse, where each instance of time travel creates a new branch of reality. In this model, the time traveller creates a new timeline in which they give the information to their younger self, rather than creating a paradoxical loop in the same timeline.

Despite its paradoxical nature, some scientists and philosophers have argued that the Bootstrap Paradox may have a real-world analog in quantum mechanics, where certain particles appear to communicate instantaneously across vast distances, without any clear source or origin. However, this interpretation remains controversial and subject to ongoing debate and investigation.

The predestination paradox

The predestination paradox, also known as the causal loop paradox, is a concept in time travel where events in the past or future are caused by events in the future or past, creating a self-contained loop of causality. This paradox challenges the concept of free will and suggests that the future is predetermined.

The predestination paradox can be illustrated by a classic example. Imagine a time traveller who goes back in time to give a famous writer a copy of their own book, which the writer then publishes under their own name, thereby making the time traveller's book a bestseller. The question arises, where did the book originally come from? If the time traveller got the book from the future and gave it to the writer, and the writer published it under their own name, then who actually wrote the book in the first place?

This paradox suggests that the time traveller's actions were predestined and that the future has already been determined. In this scenario, the time traveller's actions were necessary for the book to exist, but the book itself was necessary for the time traveller to obtain it and bring it back in time. The events create a loop where the cause and effect become indistinguishable.

The predestination paradox raises several philosophical and metaphysical questions. If the future is predetermined, then do we have free will, or are we merely actors following a predetermined script? If the future can affect the past and the past can affect the future, then what is the nature of causality and the flow of time? These questions have puzzled philosophers, scientists, and science fiction writers for decades.

One possible resolution to the predestination paradox is the idea of a multiverse, where each action creates a new reality, branching off into infinite parallel universes. In this scenario, the time traveller's actions would not necessarily create a paradox, as they would be creating a new timeline, rather than affecting the one they came from. However, this raises new questions about the nature of reality and the existence of infinite parallel universes.

Other paradoxes related to time travel and causality

Ontological paradox: A person goes back in time and changes the past in such a way that their future self is prevented from existing, which raises the question of how the person could have gone back in time in the first place.

Twin paradox: One twin travel at near the speed of light while the other stays on Earth. When the traveling twin returns, they have aged less than the twin who stayed on Earth, which seems to contradict the idea that time passes at a constant rate for all observers.

Information paradox: The paradox arising from the apparent contradiction between the laws of quantum mechanics and the laws of general relativity. In a black hole, all matter and information are thought to be destroyed, but according to quantum mechanics, information cannot be destroyed.

Fermi paradox: If the universe is so vast, why haven't we encountered any evidence of extra-terrestrial life yet?

THE PARADOXES OF TIME challenge our intuitions and encourage us to question our assumptions about the nature of time and the universe. They serve as a reminder that our everyday experiences with time may not always be a reliable guide to understanding the behaviour of time in the most extreme or unusual scenarios.

FREQUENTLY ASKED QUESTIONS ABOUT TIME

HERE LET US TRY TO question the questions we have always been asking ourselves.

I have always been in some sort of day dreaming, not trying to find solutions but just by the questions.

Does time exist independently of human perception?

Does time have a beginning or an end?

What is the relationship between time and causality?

How do we measure time?

Can time be manipulated?

How does time relate to entropy and the arrow of time?

What is the speed of time?

How does time dilation occur in special relativity?

What is the role of time in the evolution of the universe?

How do black holes affect the passage of time?

Can time run backwards?

What is the relationship between time and consciousness?

How does time impact our daily lives?

Is time an illusion or a fundamental aspect of reality?

What is the connection between time and the laws of physics?

What is the origin of time?

I really don't know whether I will be able to answer all these rightly, but just like you, I am just trying to calm down my own self by not thinking much!

Before going through the above questions, let us try some more which seem to be a bit easier!

Why does time seem to go faster as we get older?

The perception that time seems to go faster as we get older is a common experience. Many people feel that the days, weeks, and months seem to fly by faster than they did in their youth. This phenomenon is often attributed to the idea that as we age, we experience fewer new and novel experiences, and our lives become more routine and predictable. This can make time seem to pass by more quickly since we are not as aware of the passage of time.

Another explanation for this phenomenon is rooted in the concept of subjective time. According to this theory, our perception of time is based on the number of new and memorable experiences we have. When we are young, everything is new and exciting, and we are constantly learning and growing. As we age, we tend to fall into routines and familiar patterns, and we experience fewer new things. As a result, our brains perceive time as passing more quickly because we are not as aware of the passage of time.

There is also evidence to suggest that our perception of time is influenced by our emotional state. When we are happy and engaged in enjoyable activities, time seems to pass more quickly. Conversely, when we are bored or anxious, time seems to drag on. As we age, we may become more prone to negative emotions, which can make time seem to pass more slowly.

Finally, there is some evidence to suggest that the brain's processing speed slows down as we age, which can make time seem to pass more quickly. This may be due to changes in the brain's structure and function as we get older, such as a decrease in the number of neurons or changes in the levels of neurotransmitters.

While there is no clear-cut explanation for this phenomenon, it is clear that our subjective experience of time is a complex and multi-faceted aspect of human perception.

What will happen if I fall into a blackhole?

If you were to fall into a black hole, the experience would be both terrifying and incomprehensible. As you approach the event horizon, the point of no return, the gravitational forces become so intense that not even light can escape. At this point, time itself begins to behave in a strange way.

According to Einstein's theory of relativity, time slows down in the presence of strong gravitational fields. This effect, known as time dilation, becomes more pronounced as you get closer to the event horizon. From an external observer's perspective, you would appear to slow down and eventually freeze at the event horizon, never to be seen again.

However, from your perspective, time would seem to be passing normally. As you approach the event horizon, the black hole would appear to fill your entire field of view, and the universe outside would appear to be compressed into a smaller and smaller space. As you cross the event horizon, you would be stretched out into a long, thin shape known as a "spaghettification".

As you continue to fall towards the singularity at the centre of the black hole, the gravitational forces become infinitely strong, and time itself appears to come to a stop. However, from your perspective, you would continue to experience time normally until you reach the singularity, where the laws of physics as we know them break down.

In short, falling into a black hole would be a one-way journey with no hope of return. It would also provide us with valuable insights into the nature of space and time, and the limits of our understanding of the universe.

Again, your experience would even depend on the size and structure of the black hole.

For a non-rotating, static black hole, there is a point of no return called the event horizon. Once you cross this point, the gravitational force becomes so strong that it is impossible to escape. The closer you get to the black hole, the stronger the gravitational force becomes, and you would experience something called tidal forces. These forces would stretch and compress your body, causing you to become spaghettified. Eventually, you would reach the singularity at the centre of the black hole, where the gravitational forces become infinite and classical physics breaks down.

For a rotating black hole, there is also a region called the ergo sphere where space and time are dragged along with the rotation of the black hole. If you were to enter the ergo sphere, you would be forced to move in the same direction as the black hole's rotation, and it would be impossible to resist this motion. However, if you were to somehow manage to escape the ergo sphere, you could still avoid falling into the black hole by using a technique known as a slingshot manoeuvre, which involves using the black hole's gravity to accelerate yourself to a higher speed and escape its pull.

What will happen if I enter a wormhole?

The idea of a wormhole, a hypothetical shortcut through spacetime connecting two distant points, has fascinated scientists and science-fiction enthusiasts alike. While there is no empirical evidence of the existence of wormholes, theoretical physics predicts their possibility. So, if one could enter a wormhole, what might happen?

First, it's important to understand that wormholes are not like doorways or portals that allow instantaneous travel from one place to another. Rather, they are solutions to the equations of general relativity that describe how spacetime can be warped and curved by massive objects like stars and black holes. In theory, a wormhole could be created by stretching spacetime to create a "throat" or tunnel that connects two points in spacetime.

Assuming that a stable, traversable wormhole exists, what would happen to someone who enters it? One possibility is that the traveller would be subject to intense gravitational forces as they pass through the throat of the wormhole. This could cause extreme tidal forces and spaghettification, a phenomenon where the difference in gravitational pull-on different parts of the body stretches the traveller into a long, thin shape resembling a piece of spaghetti.

Another possibility is that the traveller could encounter exotic forms of matter and energy, which would be required to hold the wormhole open and prevent it from collapsing. These forms of matter and energy are purely hypothetical and have never been observed, but they are predicted by some theories of general relativity and quantum mechanics.

Assuming that the traveller survives the intense gravitational forces and encounters no exotic matter, they would emerge from the other end of the wormhole at a different point in spacetime. This could be in a different location in the same universe or in a different universe altogether, depending on the nature of the wormhole.

It's important to note that the existence of stable, traversable wormholes is purely theoretical at this point, and there is no empirical evidence to support their existence. However, the possibility of their existence raises fascinating questions about the nature of spacetime and the possibility of interstellar travel.

Will time end?

Will time end? This is a question that has puzzled humanity for centuries. Time is a fundamental aspect of our existence, and the idea that it could come to an end is a difficult concept to grasp. In this chapter, we will explore the scientific theories surrounding the end of time.

First, let's establish what we mean by the "end of time." In physics, time is often seen as an arrow that moves forward, never reversing or stopping. The idea of time ending suggests that this arrow would come to a halt, and time would cease to exist. This concept is closely related to the end of the universe, which we will discuss in more detail later.

One of the most prominent theories regarding the end of time comes from the field of cosmology. The current understanding of the universe is that it began with the Big Bang around 13.8 billion years ago. The universe has been expanding ever since, and scientists have been trying to determine the ultimate fate of this expansion.

One possible scenario is the Big Freeze. This theory suggests that the universe will continue to expand, causing galaxies to become more and more isolated from each other. Eventually, the stars in each galaxy will use up all of their fuel and die out, leaving only black holes and other dark remnants. Over an extremely long period of time, even black holes will decay, releasing particles and radiation until they too disappear. At this point, the universe will be a vast expanse of dark, empty space with no matter, energy, or life.

Another possible scenario is the Big Crunch. This theory suggests that the universe will eventually stop expanding and begin to contract. As matter becomes more densely packed, the gravitational forces will become stronger and stronger, eventually causing the entire universe to collapse in on itself in a massive implosion. This would result in a singularity, a point of infinite density and temperature. The universe would essentially be reborn, and time would start over again.

Yet another possibility is the Big Rip. This theory suggests that the expansion of the universe will accelerate to such an extent that it will eventually tear apart everything in the universe, including subatomic particles. This would happen because the repulsive force of dark energy would become stronger than the gravitational forces holding everything together.

It's important to note that these are just theories, and scientists continue to study and refine them. The ultimate fate of the universe is still a subject of much debate and speculation.

But what about time itself? Is it possible that time could end even if the universe does not? Some physicists have proposed the idea of a "timeless universe," in which time is not a fundamental aspect of reality. Instead, time is an emergent property of the universe, much like temperature is an emergent property of a collection of particles. In this view, the concept of time only exists as long as the universe exists. If the universe were to end, time would cease to exist.

Other physicists have suggested that time could come to an end even if the universe does not. This idea is based on the concept of entropy, which is a measure of the disorder or randomness in a system. According to the second law of thermodynamics, the entropy of any isolated system will always increase over time. This means that eventually, all of the matter and energy in the universe will become evenly distributed, and there will be no more gradients or differences. At this point, the universe will be in a state of maximum entropy, and time will effectively come to a halt.

However, it is important to note that these theories are still just that - theories. While they are based on scientific evidence and mathematical models, they are not definitive proof of what will happen in the future. Additionally, our understanding of time and the universe is constantly evolving, and new theories may arise that challenge or refine our current understanding.

Whether time is eternal or finite, it is a fundamental aspect of our existence and will continue to shape our lives and our understanding of the world around us.

Can time be stopped?

The question of whether time can be stopped is one that has fascinated humans for centuries. While the concept of time is a fundamental part of our understanding of the world, the nature of time itself is still not fully understood. In this chapter, we will explore the question of whether time can be stopped, drawing upon our current scientific knowledge and philosophical perspectives.

WHAT IS TIME?

Before we can address the question of whether time can be stopped, it is important to first understand what time is. Time is a concept that refers to the sequence of events that occur in the universe. It is often described as a fourth dimension, with the other three dimensions being space. Time is also intimately linked to the concept of causality, as events that occur earlier in time are said to cause events that occur later.

THE CONCEPT OF STOPPING Time

When we talk about stopping time, we are essentially talking about freezing the sequence of events that make up our experience of time. In other words, we are looking to halt the passage of time itself, so that the universe remains in a fixed state.

SCIENTIFIC PERSPECTIVES on Stopping Time

From a scientific perspective, the question of whether time can be stopped is a complex one. The laws of physics suggest that time is an irreversible process, and that it always moves forward in a continuous manner. This is known as the arrow of time, and it is intimately linked to the concept of entropy, which is the measure of the disorder or randomness of a system.

According to the second law of thermodynamics, the entropy of a closed system will always increase over time, leading to a progressive decay in order. This means that it is not possible to stop time, as any attempt to do so would violate the laws of physics. Even if we could halt the sequence of events that make up our experience of time, the underlying physical processes that govern the universe would continue to move forward.

Philosophical Perspectives on Stopping Time

From a philosophical perspective, the question of whether time can be stopped is more complex. Many philosophers argue that time is a subjective experience, and that it is our perception of time that creates the illusion of its forward motion. In this sense, stopping time would be a matter of altering our perception, rather than altering the fundamental nature of the universe.

Other philosophers argue that the question of whether time can be stopped is a meaningless one, as the very concept of time presupposes its forward motion. If we were to stop time, then we would no longer have time, and the concept itself would lose its meaning.

While our current understanding of physics suggests that time is an irreversible process that always moves forward, the philosophical debate surrounding the nature of time remains ongoing. Regardless of the answer, the concept of stopping time will likely continue to fascinate us for centuries to come.

Does time exist independently of human perception?

WHILE TIME IS A FUNDAMENTAL part of our experience of the world, it is still not fully understood whether time exists objectively, or whether it is a subjective experience that arises from human perception. In this chapter, we will explore the debate over whether time exists independently of human perception.

OBJECTIVE VS. SUBJECTIVE Time

The debate over whether time exists independently of human perception centres around the distinction between objective and subjective time. Objective time refers to time as it exists in the universe, independent of any observer. This is often thought of as the time that would exist even if there were no conscious beings to perceive it. Subjective time, on the other hand, refers to the experience of time as it is perceived by individuals. This is the time that we experience as we go about our daily lives.

SCIENTIFIC PERSPECTIVES on Objective Time

From a scientific perspective, the question of whether time exists independently of human perception is a complex one. The laws of physics suggest that time is a fundamental aspect of the universe, and that it exists independently of any conscious observer. The concept of time is intimately linked to the concept of causality, which suggests that events in the universe are connected in a temporal sequence. This connection between cause and effect requires the existence of objective time.

Furthermore, scientific theories such as relativity and quantum mechanics suggest that time is a fundamental part of the fabric of the universe. These theories provide a framework for understanding how time behaves at a fundamental level, and they suggest that time exists independently of human perception.

Philosophical Perspectives on Subjective Time

From a philosophical perspective, the question of whether time exists independently of human perception is more complex. Some philosophers argue that time is a subjective experience that arises from human perception. They suggest that our experience of time is shaped by our consciousness, and that time does not exist in the same way for non-conscious entities.

Others argue that time is a real feature of the universe, but that our perception of it is subjective. They suggest that our experience of time is shaped by the way that our brains process information, and that this processing can distort our perception of the objective reality of time.

While scientific theories suggest that time is a fundamental aspect of the universe that exists independently of any conscious observer, philosophical perspectives suggest that our perception of time is shaped by our consciousness. Ultimately, the question of whether time exists independently of human perception may be a matter of perspective, and it may be that the truth lies somewhere between these two positions.

What is the relationship between time and causality?

The relationship between time and causality is a fundamental aspect of our understanding of the world. The concept of causality refers to the relationship between an event (the cause) and a second event (the effect), where the second event is a consequence of the first. Time, on the other hand, refers to the sequence of events that occur in the universe. In this chapter, we will explore the relationship between time and causality and how they are interconnected.

CAUSALITY AND TIME in Science

In science, causality and time are intimately linked. The concept of causality is based on the idea that events that occur earlier in time can be said to cause events that occur later. This idea forms the basis for many scientific theories, and it allows scientists to make predictions about how events will unfold in the future.

In physics, the relationship between causality and time is particularly important. According to the laws of physics, causality is a fundamental principle that governs the behaviour of the universe. The theory of relativity, for example, suggests that causality is an absolute principle that holds true regardless of the observer's frame of reference.

The concept of causality is also closely linked to the concept of entropy, which is a measure of the disorder or randomness of a system. The second law of thermodynamics states that the entropy of a closed system will always increase over time, leading to a progressive decay in order. This means that the relationship between causality and time is not only important for understanding the behaviour of individual events, but also for understanding the behaviour of complex systems over time.

CAUSALITY AND TIME in Philosophy

In philosophy, the relationship between causality and time has been the subject of much debate. Many philosophers argue that the concept of causality is intimately linked to the concept of time, and that it is impossible to fully understand one without understanding the other.

Some philosophers argue that time is a necessary condition for causality, as events that occur earlier in time are understood to cause events that occur later. Others argue that the relationship between causality and time is more complex, and that causality may be an emergent property that arises from the behaviour of complex systems over time.

Still, others argue that the concept of causality is not a fundamental aspect of the universe, but rather a human construct that we use to make sense of our experience. In this view, the relationship between causality and time is not an objective feature of the universe, but rather a subjective one that arises from our perception of events.

TIME AS A SEQUENCE of Events

Time is often thought of as a sequence of events that occur in the universe. It is a fundamental aspect of our experience of the world, and it provides a framework for understanding how events are related to each other. In this sense, time provides the basis for causality, as it establishes a temporal sequence in which events occur.

CAUSALITY AS CAUSE and Effect

Causality refers to the relationship between cause and effect, and how events in the world are related to each other. It is the principle that everything that happens must have a cause, and that the effect of that cause cannot occur before the cause itself. This connection between cause and effect provides a framework for understanding how events are related to each other, and how one event can lead to another.

THE LINK BETWEEN TIME and Causality

The link between time and causality is a fundamental one. Time provides the framework for causality, as it establishes a temporal sequence in which events occur. Causality, in turn, provides a way of understanding how events in the world are related to each other, and how one event can lead to another. This connection between time and causality is essential for understanding the world around us.

CAUSALITY, TIME AND the Laws of Physics

The relationship between time and causality is also evident in the laws of physics. The laws of physics describe the behaviour of the universe, and they are intimately linked to the concept of causality. For example, the laws of motion describe how objects move through space and time, and they are based on the idea that there is a cause-and-effect relationship between the forces acting on an object and its resulting motion.

Similarly, the laws of thermodynamics describe the behaviour of energy in the universe, and they are based on the idea that there is a cause-and-effect relationship between the flow of energy and the changes that occur in a system. In both cases, the laws of physics suggest that causality is intimately linked to the passage of time, and that events that occur earlier in time are the cause of events that occur later.

CAUSALITY AND THE FLOW of Information

The relationship between time and causality is also evident in the flow of information. The flow of information refers to the way that information is transmitted from one point in space and time to another. This flow of information is intimately linked to the concept of causality, as information is often transmitted from a cause to an effect.

For example, the transmission of a signal from a radio station to a receiver is an example of the flow of information. In this case, the radio station is the cause, and the receiver is the effect. The signal is transmitted from the radio station to the receiver, and this transmission occurs in a specific sequence in time.

The concept of causality refers to the relationship between an event and a second event, where the second event is a consequence of the first. Time refers to the sequence of events that occur in the universe. The two concepts are intimately linked, and the laws of physics suggest that causality is intimately linked to the passage of time. As we continue to explore the nature of the universe, our understanding of the relationship between time and causality is likely to continue to evolve.

How do we measure time?

Time is an essential concept in our daily lives, and we use it to structure our day-to-day activities. We use time to schedule appointments, meet deadlines, and track the duration of events. But, how do we measure time?

ANCIENT METHODS FOR Measuring Time

The earliest methods for measuring time were based on natural cycles, such as the cycle of day and night, the cycle of the seasons, and the cycle of the moon. The ancient Egyptians, Greeks, and Romans used the sundial, which is the oldest known device for measuring time. A sundial works by casting a shadow onto a surface marked with hours, indicating the time of day based on the position of the sun. The sundial was accurate for measuring time during the day, but it was not effective at night or on cloudy days.

Another ancient method for measuring time is the water clock, also known as the clepsydra. The water clock was used by many civilizations, including the ancient Greeks, Persians, and Chinese. The water clock consists of two containers, one above the other, and a small hole in the bottom of the upper container. Water flows from the upper container into the lower container, and the level of water in the lower container indicates the time. The flow of water is regulated by the size of the hole and the water pressure, which affects the rate of flow. The water clock was accurate for measuring time during the day and night, but its accuracy depended on the rate of flow and the consistency of the water pressure.

MODERN-DAY METHODS for Measuring Time

Today, we have access to a wide range of methods and technologies for measuring time with greater accuracy and precision than ever before. The most widely used method for measuring time is the quartz crystal clock, which operates on the piezoelectric effect. The piezoelectric effect is a property of certain crystals that generates an electric charge when the crystal is compressed or vibrated.

In a quartz crystal clock, a small quartz crystal is vibrated at a precise frequency using an electronic oscillator. This vibration creates a precise electrical signal that can be used to measure time with an accuracy of a few parts per million. Quartz crystal clocks are used in everyday devices such as watches, clocks, and timers.

Another modern-day method for measuring time is the atomic clock. Atomic clocks are the most accurate timekeeping devices available, with an accuracy of a few parts per billion. Atomic clocks operate by measuring the vibrations of atoms, which occur at a precise frequency.

In an atomic clock, atoms are first cooled to extremely low temperatures and then trapped in a magnetic field. Lasers are then used to excite the atoms and measure their vibrations, which are used to calculate the precise time. Atomic clocks are used in scientific research, navigation systems, and communication networks.

The methods and technologies for measuring time have evolved significantly over the centuries, from the sundial and water clock to the quartz crystal clock and atomic clock. Each method has its advantages and limitations, and the choice of method depends on the desired level of accuracy and precision. Regardless of the method used, time remains a fundamental aspect of our lives, and the ability to measure it accurately and precisely is essential for scientific research, navigation, and everyday life.

How does time relate to entropy and the arrow of time?

Time, entropy, and the arrow of time are all intimately connected concepts in physics. Understanding their relationship is crucial for understanding many phenomena in the natural world, including the origins of the universe, the behaviour of complex systems, and the nature of causality.

Entropy is a measure of the disorder or randomness of a system. The second law of thermodynamics states that in a closed system, the total entropy always increases over time. This means that as time passes, the system becomes more disordered and random. For example, a cup of hot coffee left on a table will eventually cool down to room temperature, because the heat energy will dissipate into the surrounding environment, increasing the total entropy of the system.

The arrow of time refers to the directionality of time. It is a fundamental property of the universe that time flows in a particular direction, from the past to the future. The arrow of time is intimately connected to the second law of thermodynamics, because the direction of time can be inferred from the direction in which entropy is increasing. In other words, the arrow of time points in the direction of increasing entropy.

This relationship between time and entropy is fundamental to many physical processes. For example, the behaviour of stars and galaxies is strongly influenced by the flow of energy and matter through space, which is driven by the increase in entropy. The origin and evolution of the universe itself can also be explained in terms of the relationship between time, entropy, and the arrow of time. According to the big bang theory, the universe began in a highly ordered state, with very low entropy. As the universe expanded and cooled, the entropy increased, driving the formation of stars, galaxies, and eventually life itself.

The connection between time, entropy, and the arrow of time is also crucial for understanding the nature of causality. In a closed system, events occur in a particular order because of the constraints imposed by the arrow of time and the increase in entropy. For example, we can be confident that the cup of hot coffee did not spontaneously heat up because we know that heat always flows from hot to cold, in accordance with the second law of thermodynamics.

However, this relationship between time, entropy, and the arrow of time raises some interesting philosophical questions. For example, if the direction of time is determined by the increase in entropy, could there be a universe in which entropy decreases over time, and the arrow of time points in the opposite direction? And if the arrow of time is a fundamental property of the universe, what is the ultimate cause of this directionality?

The second law of thermodynamics tells us that the total entropy of a closed system always increases over time, and this increase in entropy determines the direction of the arrow of time. By understanding the connection between time, entropy, and the arrow of time, we can gain a deeper understanding of the workings of the natural world, and the fundamental properties of the universe itself.

One interesting consequence of the relationship between time, entropy, and the arrow of time is that it implies that certain physical processes are irreversible. For example, if you drop a glass on the floor and it shatters, you can't reverse the process and un-shatter the glass. This is because the process of shattering the glass increases the entropy of the system, and the direction of time is determined by the increase in entropy. If you were able to reverse the process and put the shattered glass back together, you would be decreasing the entropy of the system, which would violate the second law of thermodynamics.

The irreversibility of certain physical processes is related to the concept of the "arrow of time asymmetry." This refers to the fact that certain processes, like the shattering of a glass or the burning of a piece of paper, only occur in one direction of time. They are irreversible because the direction of time is determined by the increase in entropy, which can only occur in one direction. However, some processes, like the behaviour of subatomic particles, are reversible and do not exhibit the arrow of time asymmetry.

The concept of time and the arrow of time is also important in the field of cosmology, the study of the origins and evolution of the universe. The arrow of time is intimately connected to the origin of the universe and the idea of the "initial low entropy state." The initial low entropy state refers to the highly ordered and low entropy state of the universe at the beginning of time, and is thought to be the starting point of the increase in entropy that we observe today.

The question of why the initial low entropy state existed is still an open question in cosmology, and is often referred to as the "cosmological arrow of time problem." Some physicists have suggested that the initial low entropy state could be explained by a hypothetical process called "cosmic inflation," which would have caused the universe to expand rapidly and evenly, resulting in a highly ordered initial state.

The Arrow of Time

The Arrow of Time is a concept that refers to the asymmetry of time, and the observation that certain physical processes only occur in one direction. The arrow of time can be understood as the direction in which the natural world appears to evolve over time, and is intimately connected to the concept of entropy.

Entropy is a measure of the disorder or randomness of a system, and is related to the number of possible ways in which the system can be arranged. The second law of thermodynamics states that the total entropy of a closed system always increases over time. This means that as time passes, the universe becomes more disordered and chaotic.

The arrow of time is related to the increase in entropy, because the direction of time is determined by the direction of the increase in entropy. In other words, as time passes, entropy always increases, and this increase in entropy determines the direction of time. This is why certain physical processes, such as the shattering of a glass, only occur in one direction of time. They are irreversible because the direction of time is determined by the increase in entropy, which can only occur in one direction.

The concept of the arrow of time is important in many areas of physics, including thermodynamics, statistical mechanics, and cosmology. In thermodynamics, the arrow of time is related to the second law of thermodynamics, which is one of the fundamental laws of nature. The second law states that the total entropy of a closed system always increases over time, which means that the direction of time is always from low entropy to high entropy.

In statistical mechanics, the arrow of time is related to the concept of time-reversal symmetry. Time-reversal symmetry is the idea that if you reversed the direction of time, the laws of physics would still be valid. However, the arrow of time asymmetry means that certain physical processes are irreversible, and would not occur in a reversed direction of time.

In cosmology, the arrow of time is related to the origin of the universe and the idea of the "initial low entropy state." The initial low entropy state refers to the highly ordered and low entropy state of the universe at the beginning of time, and is thought to be the starting point of the increase in entropy that we observe today. The question of why the initial low entropy state existed is still an open question in cosmology, and is often referred to as the "cosmological arrow of time problem."

One of the interesting aspects of the arrow of time is that it is closely tied to the subjective experience of time. Our perception of time is linked to the direction of the arrow of time, and we experience time as moving in a specific direction. We remember the past, but we cannot remember the future, and we always experience the present moment as moving forward into the future.

This subjective experience of time is closely tied to the idea of causality. Causality is the relationship between an event (the cause) and a second event (the effect), where the second event is a result of the first. The arrow of time is intimately connected to causality, because the direction of time is determined by the causal relationships between events.

For example, if you drop a glass and it shatters, the direction of time is determined by the causal relationship between dropping the glass and it shattering. If you were to reverse the direction of time, the glass would magically reassemble itself and fly back up into your hand, which is not what we observe in the natural world.

The arrow of time is also related to the concept of memory. We remember the past, but we cannot remember the future. This is because memories are tied to the increase in entropy, and the direction of time is determined by the increase in entropy. When we remember something, we are retrieving information from the past, which means we are accessing a lower entropy state.

One of the key implications of the arrow of time is the irreversibility of certain physical processes. While the laws of physics are time-symmetric, meaning they are valid in both forward and backward directions of time, some processes only occur in one direction. For example, a cup of hot coffee will gradually cool down and become lukewarm, but it will not spontaneously heat up again to become hot. This is because the process of cooling down involves an increase in entropy, whereas the process of heating up would require a decrease in entropy, which is statistically unlikely.

The irreversibility of certain processes is closely linked to the concept of the arrow of time, and has implications for a range of phenomena, including biological systems and the evolution of the universe. In biological systems, the arrow of time is linked to the process of aging. As living organisms age, their cells accumulate damage and entropy increases, eventually leading to death. While the mechanisms behind aging are still not fully understood, the arrow of time is thought to play a role in this process.

In the study of the universe, the arrow of time is linked to the process of cosmic evolution. The universe has been evolving and expanding since the Big Bang, and the direction of this evolution is determined by the arrow of time. The universe is currently in a state of high entropy, with galaxies moving farther apart and becoming increasingly disordered. However, the initial state of the universe after the Big Bang was one of low entropy, which raises questions about the directionality of time and the origins of the universe.

One of the most intriguing aspects of the arrow of time is its relationship to the concept of free will. Some philosophers have argued that the directionality of time implies that we have free will, as our actions can have irreversible consequences. This idea has been debated by many scholars, but the relationship between the arrow of time and free will remains an open question.

Time and the Universe

THE CONCEPT OF TIME is closely linked to our understanding of the universe, as it plays a fundamental role in the way we measure and describe physical phenomena. From the evolution of galaxies to the behaviour of subatomic particles, time is a critical dimension that allows us to make sense of the universe and its workings.

One of the most intriguing aspects of time and the universe is the relationship between the two. As the universe has evolved and expanded over billions of years, the concept of time has played a key role in shaping its history and structure.

The origins of time in the universe can be traced back to the Big Bang, the event that is thought to have marked the beginning of the universe as we know it. According to current models of cosmology, the Big Bang occurred approximately 13.8 billion years ago and marked the birth of the universe. In the early stages of the universe, time was tightly linked to space, forming a single four-dimensional fabric known as spacetime.

As the universe expanded and cooled, the fabric of spacetime began to stretch and change, leading to the formation of galaxies, stars, and other structures. The behaviour of these structures is governed by the laws of physics, which provide a framework for understanding the behaviour of matter and energy over time.

One of the key concepts that arise from the study of the universe and time is the arrow of time. This concept describes the directionality of time, which is closely linked to the second law of thermodynamics and the concept of entropy. In simple terms, the arrow of time describes the fact that certain physical processes only occur in one direction, from lower to higher entropy, and not in reverse.

The arrow of time has important implications for our understanding of the universe, as it helps to explain why certain phenomena occur in the way they do. For example, the formation of stars and galaxies is a process that is driven by the increase in entropy as matter clumps together under the influence of gravity. This process is irreversible and is linked to the direction of time.

Another key concept that is linked to time and the universe is the expansion of the universe. The universe has been expanding since the Big Bang, with galaxies moving away from each other at an accelerating rate. This expansion is driven by dark energy, a mysterious force that is thought to permeate the universe and exert a repulsive force on matter.

The expansion of the universe has important implications for our understanding of time and the evolution of the universe. As the universe expands, the fabric of spacetime is stretched, leading to a slowing down of time in regions of high gravity. This effect, known as time dilation, is observed in the behaviour of objects in strong gravitational fields, such as black holes.

The relationship between time and the universe is complex and multi-faceted, and there are many other aspects of this relationship that are worth exploring.

For example, the concept of time is closely linked to the concept of causality, or the idea that events have causes and effects that occur in a specific order. In physics, causality is often described using the concept of time asymmetry, which states that the cause of an event always occurs before the effect.

The relationship between time and causality has important implications for our understanding of the universe, as it helps us to make sense of the behaviour of matter and energy over time. For example, the behaviour of particles in the quantum world is often described using probabilistic models that take into account the probability of different outcomes occurring over time.

Another important aspect of the relationship between time and the universe is the role that time plays in the evolution of life. The concept of biological time describes the way that living organisms experience time, and the ways in which their behaviour is shaped by the passage of time.

For example, circadian rhythms are a type of biological clock that help to regulate many physiological processes in organisms, including sleep-wake cycles, hormone production, and metabolic activity. These rhythms are driven by a combination of internal biological processes and external cues, such as the cycle of light and dark in the environment.

Time is a fundamental part of our subjective experience of the world, and it shapes our perception of events and our sense of identity over time.

For example, memories are a way in which we create a sense of continuity and narrative coherence over time, allowing us to construct a sense of self that persists over time. Similarly, the experience of time can be altered by external factors, such as drugs or extreme environments, leading to altered states of consciousness or changes in perception.

From the origins of time in the Big Bang to the behaviour of matter and energy over billions of years, time provides a critical dimension that allows us to make sense of the workings of the universe. The arrow of time and the expansion of the universe are just two of the many concepts that are linked to time and the universe, and researchers continue to explore these and other topics in order to deepen our understanding of the universe and our place within it.

The relationship between time and entropy has significant implications for the behaviour of the universe as a whole. According to the Big Bang Theory, the universe began as an extremely low-entropy state, with all matter and energy concentrated in a single point. As the universe expanded and cooled, it began to move towards a state of maximum entropy, with matter and energy dispersed throughout space. This process has continued over billions of years, resulting in the universe we see today.

The role of time in this process cannot be overstated. Without the arrow of time and the irreversible increase in entropy, the universe could not have evolved from a low-entropy state to a high-entropy state. The flow of time provides a directionality to the behaviour of the universe, allowing it to evolve and change over time.

The increase in entropy over time provides a directionality to the flow of time, known as the arrow of time, and plays a crucial role in the behaviour of the universe as a whole. Understanding this relationship is essential for understanding the behaviour of complex systems, such as the universe at large scales, and has practical applications in fields such as thermodynamics, cosmology, and information theory.

What is the speed of time?

The concept of the speed of time can be a somewhat confusing one, as time is not a physical object that can be measured in the same way as other quantities such as length or mass. However, there are certain aspects of time that can be quantified and compared, which can provide insight into how time operates and how it relates to our perception of the world around us.

TIME AS A CONSTRUCT

First, it is important to understand that time is a construct that we use to make sense of the world around us. We measure time in units such as seconds, minutes, and hours, and use it to order events and make predictions about the future. However, time itself is not a physical object that can be touched or measured directly. Rather, it is a concept that we use to describe the progression of events in the world.

THE SPEED OF TIME PERCEPTION

One way in which the concept of the speed of time is often discussed is in relation to our perception of time. We all experience time in different ways, and some situations may cause time to seem to move faster or slower than others. For example, time may seem to move more slowly when we are bored or in pain, while it may seem to pass more quickly when we are engaged in a fun or interesting activity.

Research has shown that our perception of time can be influenced by a number of factors, including our age, the amount of attention we are paying to our surroundings, and the complexity of the stimuli we are experiencing. For example, children often perceive time to move more slowly than adults, as their brains are still developing and they are experiencing many things for the first time.

TIME AND THE SPEED of Light

Another way in which the speed of time is often discussed is in relation to the speed of light. According to the theory of relativity, the speed of light is the fastest speed at which anything can travel through space. This has significant implications for the way we perceive time, as objects that are moving at high speeds relative to one another will experience time differently.

For example, if a person were to travel away from Earth at a significant fraction of the speed of light and then return, they would find that less time had passed for them than for someone who remained on Earth. This is known as time dilation, and it is a fundamental aspect of the theory of relativity.

Our perception of time can be influenced by a number of factors, and the way in which we experience time can have significant implications for our lives. By understanding the ways in which time operates and the factors that influence our perception of it, we can gain a greater appreciation for the complexity and beauty of the world around us.

How does time dilation occur in special relativity?

In special relativity, time dilation is a phenomenon that occurs when two observers moving relative to each other measure different amounts of time between two events. This effect arises due to the fundamental postulate of relativity that the speed of light is constant in all reference frames, and has been experimentally verified in numerous experiments.

To understand how time dilation occurs, we need to first establish some basic concepts from special relativity. One of the key ideas in special relativity is that space and time are intimately connected, and can be thought of as different aspects of a single entity known as spacetime. Spacetime can be thought of as a four-dimensional grid that contains all the events that have ever occurred or will ever occur.

Another important concept is the principle of relativity, which states that the laws of physics are the same for all observers in uniform motion relative to each other. This means that there is no preferred reference frame in the universe, and that all physical measurements must be made relative to some observer.

Now, consider two observers, A and B, who are moving relative to each other with a constant velocity. Suppose that they both observe a clock that is located at rest in their respective reference frames, and that the clock starts ticking at the same time in both frames.

According to special relativity, the speed of light is the same in both frames, regardless of the motion of the observer. This means that the time it takes for light to travel between the clock and the observer will be different for observer A and observer B, since they are moving relative to each other.

As a result, observer A and observer B will measure different amounts of time between the start and end of the clock ticking. This effect is known as time dilation, and it arises because the relative motion between the two observers causes their measurement of time to become stretched or compressed.

The amount of time dilation that occurs depends on the relative velocity between the two observers, and can be calculated using the Lorentz transformation equations. These equations describe how measurements of time and space are related in different reference frames, and they are an essential tool for understanding the effects of special relativity.

In general, the faster two observers are moving relative to each other, the more time dilation will occur. At speeds approaching the speed of light, time dilation becomes extreme, and can result in apparent time travel and other bizarre effects.

One interesting consequence of time dilation is that it can be used to explain the phenomenon of cosmic rays, which are high-energy particles that are constantly bombarding the Earth from outer space. These particles are thought to be produced by distant supernovae and other astrophysical phenomena, and they are able to reach the Earth because they are traveling at very high speeds relative to us.

Because of their high speeds, cosmic rays experience significant time dilation as they travel through space, and this can cause them to appear to have much longer lifetimes than they would if they were at rest. This effect allows cosmic rays to travel much farther than they would otherwise be able to, and it has important implications for our understanding of the universe.

Time dilation is a fundamental consequence of special relativity, and arises due to the fact that the speed of light is constant in all reference frames. This effect leads to differences in the measurement of time between observers who are moving relative to each other, and has been experimentally verified in numerous experiments. The concept of time dilation is an important part of our understanding of the universe, and has important implications for our understanding of cosmology, particle physics, and other fields.

What is the role of time in the evolution of the universe?

The role of time in the evolution of the universe is a fundamental aspect of modern cosmology. Time is a crucial element that governs the behaviour of matter and energy in the universe, and it plays a critical role in determining the evolution and fate of the cosmos.

From the moment of the Big Bang, time has been an essential component of the universe. As the universe began to expand, time started ticking, and this process has continued ever since. The expansion of the universe is driven by the forces of gravity and the properties of matter and energy, all of which are influenced by the passage of time.

In the early moments of the universe, time played a vital role in determining the structure and composition of the cosmos. During the first few microseconds, the universe was filled with a hot, dense plasma of particles, which were constantly interacting with each other. As time passed, the universe began to cool and expand, allowing matter to form into more complex structures, such as atoms, stars, and galaxies.

Over billions of years, the evolution of the universe has been driven by the forces of gravity and the properties of matter and energy. These forces have shaped the structure of the cosmos, determining the formation of galaxies, clusters of galaxies, and the vast cosmic web that connects them.

The passage of time has also played a critical role in the evolution of stars, which are the engines that power the cosmos. As stars age, they undergo a series of changes, eventually leading to their ultimate fate as either a black hole or a white dwarf. The amount of time that a star spends on the main sequence, where it burns hydrogen into helium, is determined by its mass, with more massive stars having shorter lifetimes than less massive ones.

Time also plays a crucial role in the evolution of life on Earth. The history of life on our planet is a story of evolution, adaptation, and extinction, all of which have been driven by the passage of time. The slow march of time has allowed living organisms to adapt to changing environments, leading to the incredible diversity of life that we see around us today.

In cosmology, time is intimately connected with the concept of cosmic expansion, which is one of the most fundamental aspects of modern cosmology. The expansion of the universe is an ongoing process, and it is the key to understanding many of the phenomena that we observe in the cosmos, such as the redshift of distant galaxies and the cosmic microwave background radiation.

The role of time in the evolution of the universe is essential and far-reaching. It plays a critical role in shaping the structure and composition of the cosmos, determining the evolution of stars and galaxies, and driving the evolution of life on Earth. Time is intimately connected with the concept of cosmic expansion, and it has played a vital role in our understanding of the cosmos since the dawn of modern cosmology.

How do black holes affect the passage of time?

Black holes are one of the most fascinating and mysterious objects in the universe. They are formed when massive stars run out of fuel and collapse under the force of their own gravity, becoming so dense that nothing, not even light, can escape their grasp. One of the most intriguing properties of black holes is their ability to affect the passage of time. In this article, we will explore how black holes affect the flow of time and what this means for our understanding of the universe.

GENERAL RELATIVITY and Time Dilation

The concept of time dilation, or the slowing down of time under certain conditions, is a fundamental prediction of Albert Einstein's theory of general relativity. According to this theory, gravity is not a force that acts between two objects, as in Newton's theory of gravity, but is instead the curvature of spacetime caused by the presence of matter and energy. This curvature can cause time to pass more slowly in regions of strong gravity, a phenomenon known as gravitational time dilation.

BLACK HOLES AND TIME Dilation

Black holes are the ultimate expression of strong gravity, and their intense gravitational fields can cause extreme time dilation effects. As an object falls towards a black hole, it experiences increasing gravitational forces, and time appears to slow down from an observer's perspective. If an object were to get close enough to the event horizon, the point of no return, time would appear to stop altogether, and the object would be frozen in time.

This effect is known as gravitational time dilation, and it is one of the most profound consequences of general relativity. The closer an object gets to a black hole, the more time slows down, and the more extreme the effects become. This means that for an observer far from a black hole, time would appear to be moving at a different rate than for an observer close to the black hole.

BLACK HOLES AND TIME *Travel*

The extreme time dilation effects of black holes have led to speculation about the possibility of time travel. In theory, if an object were to get close enough to a black hole, it could experience time dilation so extreme that it could travel forward in time relative to the rest of the universe.

However, the practicalities of this are far from straightforward. For one thing, an object would have to get extremely close to a black hole to experience significant time dilation, which would be incredibly dangerous due to the black hole's intense gravitational forces. Additionally, the object would have to find a way to escape the black hole's gravitational pull, which is notoriously difficult.

IMPLICATIONS FOR OUR *Understanding of the Universe*

The extreme time dilation effects of black holes have profound implications for our understanding of the universe. They suggest that time is not a universal constant but is instead relative to the observer's perspective and the strength of the gravitational field they are in.

Furthermore, the effects of black holes on time may have important consequences for our understanding of the evolution of the universe. Black holes are thought to play a significant role in the formation and evolution of galaxies, and their effects on time may have helped shape the structure of the universe as we know it today.

Their ability to affect the flow of time is one of the most profound consequences of general relativity and has important implications for our understanding of the universe. While the extreme time dilation effects of black holes may never allow us to travel through time, they have helped us gain a deeper appreciation for the fundamental nature of time and its relationship to gravity.

Can time run backwards?

The question of whether time can run backwards is a fascinating one that has puzzled philosophers and scientists for centuries. While the concept of time travel has been popularized in science fiction, the reality is that according to our current understanding of physics, time always moves forward and cannot run backwards.

The arrow of time, which describes the one-way direction of time from past to future, is a fundamental concept in physics. It is closely related to the second law of thermodynamics, which states that the entropy of a closed system tends to increase over time. Entropy is a measure of disorder or randomness, and the second law implies that the future of any system is more uncertain than its past, and the past cannot be fully reconstructed from the present.

This concept of irreversibility is a crucial part of our understanding of the physical world. It is also evident in our everyday experience, as we observe events unfolding in a particular sequence, and we cannot change the past or predict the future with certainty.

Despite this, there are some speculative theories in physics that suggest the arrow of time may not be an absolute rule. For example, certain interpretations of quantum mechanics suggest that the arrow of time may be an emergent property of the universe, rather than a fundamental law. Additionally, the possibility of time travel raises the question of whether it is possible to go back in time and change the past, thus reversing the arrow of time.

The question of whether time can run backwards is not only a scientific one but also a philosophical and metaphysical one. It raises fundamental questions about the nature of time, causality, and the directionality of the universe.

In addition to the concept of irreversibility, the arrow of time is also closely related to the concept of causality. The idea that events have causes and effects, and that the effect cannot precede the cause, is intimately linked with the concept of the arrow of time. If time could run backwards, then causality would be reversed as well, and events could occur before their causes, which would create logical contradictions.

The directionality of time is closely related to the expansion of the universe. According to the Big Bang theory, the universe began as a singularity and has been expanding ever since. This expansion creates a sense of time moving forward, as the universe becomes more disordered and the entropy increases. If time could run backwards, then the universe would have to contract, which would contradict our observations of the universe's expansion.

There are still some unresolved questions in physics that suggest the possibility of time reversal. For example, some studies suggest that certain subatomic particles may violate the arrow of time and move backward in time, although this is still a subject of debate and investigation.

However, these theories remain highly speculative and are not yet supported by empirical evidence. For practical purposes, we can say that time cannot run backwards in our everyday experience, and the arrow of time remains an important concept in physics and philosophy alike.

What is the relationship between time and consciousness?

The relationship between time and consciousness has been a topic of philosophical and scientific inquiry for centuries. While the concept of time is often viewed as a universal constant, the experience of time can vary widely between individuals, leading to questions about how time is perceived and processed by the brain.

At its core, consciousness refers to the subjective experience of being aware of oneself and one's surroundings. This experience is intimately linked to the perception of time, as individuals use their awareness of time to orient themselves in their surroundings and make sense of the world around them.

One of the most fundamental aspects of time perception is its malleability. While time is often thought of as a linear progression, it can be experienced in a variety of ways depending on factors such as age, mood, and attentional state. For example, time can seem to pass more quickly when we are engaged in an enjoyable activity or when we are focused on a task, while it can drag on endlessly when we are bored or waiting for something to happen.

The processing of time in the brain is a complex process that involves multiple areas of the brain, including the prefrontal cortex, the parietal cortex, and the basal ganglia. These areas work together to create a sense of temporal coherence, allowing us to perceive the world around us as a continuous flow of events.

One of the most intriguing aspects of time perception is its potential relationship to consciousness itself. Some researchers have suggested that consciousness may be an emergent property of the brain's ability to process and integrate information over time. In this view, the brain creates a sense of consciousness by constantly integrating sensory information and updating our mental models of the world in real-time.

However, the relationship between time and consciousness is far from fully understood. While there is evidence to suggest that the brain's processing of time is intimately linked to our subjective experience of consciousness, the precise nature of this relationship remains a topic of ongoing research.

One particularly fascinating avenue of research in this area is the study of altered states of consciousness. Researchers have found that experiences such as meditation, psychedelic drug use, and near-death experiences can all lead to profound changes in the perception of time. These experiences have the potential to shed light on the nature of consciousness itself and the relationship between time and subjective experience.

One of the main theories about the relationship between time and consciousness is the "internal clock" hypothesis, which suggests that the brain has an internal mechanism for measuring time. This mechanism is thought to be based on the firing patterns of neurons in the brain, which can be used to create a sense of temporal order and duration.

Research has shown that different areas of the brain are involved in processing different aspects of time perception. For example, the prefrontal cortex is thought to be involved in temporal sequencing, while the parietal cortex is involved in spatial-temporal integration.

The subjective experience of time can also be influenced by external factors such as social context and cultural norms. For example, different cultures may have different concepts of time, with some emphasizing punctuality and efficiency while others prioritize a more relaxed, leisurely pace of life.

One of the most intriguing aspects of the relationship between time and consciousness is its potential relevance to the question of free will. If our sense of consciousness is based on the brain's ability to process information over time, it raises the question of whether our actions are truly under our control or whether they are predetermined by the laws of physics.

Some philosophers and scientists have suggested that our sense of free will is an illusion created by our subjective experience of time. In this view, our conscious experience of choice and agency is merely a by-product of the brain's internal processing, and our actions are ultimately determined by factors beyond our conscious control

While much progress has been made in understanding how the brain processes and perceives time, there is still much to be learned about the relationship between time and consciousness. Future research in this area has the potential to shed light on some of the most fundamental questions about the nature of consciousness itself.

How does time impact our daily lives?

The way we perceive time can also vary greatly depending on our individual circumstances and life experiences. For example, people who live in fast-paced urban environments may feel like time is constantly slipping away, while those living in more rural areas may have a more relaxed and leisurely perception of time.

Moreover, the way we use our time can have a profound impact on our overall quality of life. When we spend our time engaged in activities that we enjoy and find fulfilling, we tend to feel happier and more satisfied with our lives. Conversely, when we spend our time in activities that we find tedious, stressful, or unfulfilling, we may feel unhappy, stressed, and unfulfilled.

One of the most significant ways that time impacts our daily lives is through the aging process. As we age, our perception of time may change, with days and weeks seeming to pass more quickly than they did in our youth. This can lead to a sense of urgency and a desire to make the most of our remaining time, as well as feelings of nostalgia for past experiences and a sense of regret for missed opportunities.

One of the most significant ways that time impacts our daily lives is through our work schedules and routines. Most people work a set number of hours per day or week, and these schedules are often tightly structured and inflexible. As a result, many people feel a constant sense of time pressure and urgency, as they try to balance their work responsibilities with other commitments and obligations.

Time also plays a crucial role in our personal relationships and social interactions. We use time to plan and coordinate activities with friends and family, and to mark important milestones and events in our lives. From birthdays and anniversaries to graduations and weddings, time provides a framework for celebrating and commemorating important moments in our lives.

Another way that time impacts our daily lives is through our physical and mental health. Sleep patterns, meal schedules, and exercise routines are all strongly influenced by the body's internal clock, or circadian rhythm. Disruptions to these rhythms can lead to a variety of health problems, including insomnia, depression, and weight gain.

The way we experience time can also have a profound impact on our emotional well-being. When we feel like time is passing quickly, we may feel a sense of urgency or pressure to accomplish our goals and meet our deadlines. Conversely, when time seems to drag on, we may feel bored, restless, or frustrated.

The pace of technological change is another factor that impacts our daily lives. The rapid development of new technologies and digital platforms has made it easier than ever to stay connected and informed, but it has also created new pressures and demands on our time. Many people feel overwhelmed by the constant stream of emails, notifications, and messages that they receive on a daily basis, leading to a sense of information overload and burnout.

Understanding the ways in which time shapes our experiences and perceptions can help us to better manage our time, prioritize our goals, and live happier, more fulfilling lives.

Is time an illusion or a fundamental aspect of reality?

Is time an illusion, a mere human construct that we use to organize our experiences, or is it a fundamental aspect of reality, an objective feature of the universe? This is a question that has been debated by scholars and scientists for centuries, and while there is no easy answer, there are compelling arguments on both sides.

One of the most persuasive arguments for the idea that time is an illusion is based on the fact that time is a subjective experience. We perceive time differently depending on our individual circumstances, and it seems to be shaped by our perceptions, emotions, and experiences. For example, when we are bored or anxious, time may seem to pass more slowly, while when we are engaged in enjoyable activities, time may seem to fly by. This variability in our experience of time suggests that it may be more of a human construct than an objective feature of the universe.

Moreover, the concept of time itself is difficult to define and understand. Philosophers have long debated the nature of time, and there is still no consensus on what it is or how it works. Some argue that time is a fundamental aspect of reality, like space or matter, while others view it as a mere abstraction, a tool that we use to organize our experiences.

However, there are also compelling arguments for the idea that time is a fundamental aspect of reality. One of the most powerful arguments in favour of this view is based on the fact that time seems to be inexorably linked to the physical universe. We see evidence of time in the motion of planets and stars, in the decay of radioactive elements, and in the biological rhythms of living organisms. This suggests that time may be an objective feature of the universe, one that is intimately connected to the workings of the physical world.

The laws of physics themselves seem to require the existence of time. Einstein's theory of relativity, for example, shows that time is inseparable from space, and that the two are intimately connected in a four-dimensional fabric known as spacetime. This implies that time is a necessary component of the physical universe, one that cannot be reduced to a mere human construct or illusion.

Those who argue that time is an illusion believe that it is simply a human construct, a way for us to measure and organize the sequence of events that occur in our lives. According to this view, time is not a physical entity that exists independently of our perception, but rather a concept that we have created to help us make sense of the world around us.

Proponents of this view often point to the fact that time can be experienced in different ways, depending on our perspective and circumstances. For example, time may seem to pass more slowly when we are bored or anxious, while it may seem to fly by when we are engaged in enjoyable activities. This variability suggests that time is not a fixed, objective phenomenon, but rather something that is shaped by our perception and experience.

On the other hand, many scientists and philosophers argue that time is a fundamental aspect of reality, an objective feature of the universe that exists independently of human perception. According to this view, time is not just a human construct, but rather a fundamental aspect of the fabric of the universe, a dimension that exists alongside space and matter.

Evidence for this perspective can be found in numerous areas of science, from Einstein's theory of relativity to quantum mechanics. In these fields, time is treated as a fundamental component of the universe, with physical laws and phenomena that are closely tied to the passage of time.

Moreover, time plays a crucial role in our understanding of causality and the nature of change. Without the concept of time, it would be impossible to make sense of the way in which events unfold in the universe or to understand the relationship between cause and effect.

Despite these arguments, the question of whether time is an illusion or a fundamental aspect of reality remains unresolved. While some argue that time is merely a human construct, others maintain that it is a fundamental dimension of the universe, inseparable from space and matter.

The debate over whether time is an illusion or a fundamental aspect of reality is a complex and ongoing one. While there are persuasive arguments on both sides, it seems likely that time is a complex and multifaceted concept that defies easy definition or explanation. Whether it is an objective feature of the universe or a human construct, time remains an essential element of our lives, shaping our experiences, perceptions, and interactions with the world around us.

What is the connection between time and the laws of physics?

Time is a fundamental concept that permeates our lives, and it is intimately connected to the laws of physics. The laws of physics describe the behaviour of the natural world, and time is a critical aspect of that behaviour. In this article, we will explore the connection between time and the laws of physics and examine how this connection has shaped our understanding of the universe.

TIME IN PHYSICS

In physics, time is defined as the progression of events from the past through the present and into the future. It is a fundamental component of the universe, and it plays a critical role in our understanding of the physical world. Time is typically measured in units such as seconds, minutes, hours, days, and years.

One of the most fundamental laws of physics is the law of cause and effect, which states that every action has an equal and opposite reaction. This law requires the passage of time, as cause and effect must occur in a sequence. The law of cause and effect is intimately connected to the concept of time, as it requires a progression from cause to effect.

The laws of physics are a set of fundamental principles that describe how the physical world behaves. They are based on observations and experiments, and they provide a framework for understanding and predicting the behaviour of the natural world. Time is an essential aspect of physics, and it plays a critical role in the laws of physics.

The laws of physics are formulated in a way that describes how time influences the behaviour of the natural world. One of the most fundamental laws of physics is the law of cause and effect, which states that every action has an equal and opposite reaction. This law requires the passage of time, as cause and effect must occur in a sequence.

The laws of motion describe how the position, velocity, and acceleration of an object change over time. According to Newton's first law of motion, an object at rest will remain at rest unless acted upon by an external force, while an object in motion will remain in motion at a constant velocity unless acted upon by an external force. The second law of motion states that the force acting on an object is proportional to its mass and acceleration, while the third law of motion states that every action has an equal and opposite reaction. These laws require the passage of time, as an object's position, velocity, and acceleration change over time.

The laws of thermodynamics describe how the energy of a system changes over time. The first law of thermodynamics, also known as the law of conservation of energy, states that energy cannot be created or destroyed, only transferred or converted from one form to another. The second law of thermodynamics states that the entropy, or disorder, of a closed system will always increase over time. These laws require the passage of time, as the energy of a system changes over time.

The laws of electromagnetism describe how electric and magnetic fields change over time. Maxwell's equations describe how electric and magnetic fields interact with each other and with charged particles, and they provide a framework for understanding phenomena such as electromagnetic waves and radiation. These laws require the passage of time, as electric and magnetic fields change over time.

The theory of relativity, proposed by Albert Einstein in 1905, revolutionized our understanding of time and its relationship to the laws of physics. According to the theory of relativity, time is not an absolute concept, but rather, it is relative to the observer's frame of reference. This means that time can appear to pass differently for different observers, depending on their relative motion and position in space. For example, if two observers are moving at different speeds, they will experience time differently. One observer may perceive time as passing more slowly than the other observer.

The concept of time dilation is a crucial aspect of the theory of relativity. Time dilation occurs when time appears to pass more slowly for an object that is moving at high speeds. This phenomenon has been observed in numerous experiments and has been verified through the use of highly accurate atomic clocks.

The laws of physics are formulated in a way that describes how time influences the behaviour of the natural world, and they provide a framework for understanding and predicting the behaviour of the physical world. The theory of relativity has revolutionized our understanding of time, revealing that time is not an absolute concept but is relative to the observer's frame of reference. The laws of physics and time are intimately connected, and our understanding of one informs our understanding of the other.

Chemistry without the concept of time

Time is an essential aspect of chemistry, as chemical reactions and processes occur over time. However, to imagine chemistry without the concept of time, we must first understand what it means to remove time from the equations.

In a world without time, all events would happen simultaneously, and there would be no cause-and-effect relationships. Every event would occur at the same moment, and there would be no way to differentiate between past, present, and future. In such a world, there would be no way to study chemistry, as chemical reactions and processes require time to occur.

Chemical reactions involve the rearrangement of atoms and molecules to form new substances. These reactions take place over a specific period, and the rate of the reaction depends on several factors such as temperature, pressure, and concentration. Without the concept of time, these factors would be meaningless, and there would be no way to determine the rate of a chemical reaction.

For example, let us consider the reaction between hydrogen and oxygen to form water. This reaction is exothermic, meaning it releases energy in the form of heat. The reaction proceeds according to the equation:

$$2H_2(g) + O_2(g) \rightarrow 2H_2O(l) + energy$$

Without the concept of time, we would not be able to determine how long it takes for this reaction to occur. We would not be able to measure the energy released during the reaction or study the intermediates formed during the reaction. In other words, chemistry would not exist without the concept of time.

Moreover, chemical reactions are also influenced by factors such as pressure and concentration. Pressure is a measure of the force exerted by a gas, while concentration is a measure of the number of particles in a solution. Without the concept of time, these factors would be meaningless, and there would be no way to study their effect on chemical reactions.

In addition, the concept of equilibrium is a crucial aspect of chemistry. Chemical equilibrium occurs when the rates of the forward and reverse reactions are equal, and the concentrations of the reactants and products remain constant. The position of the equilibrium depends on several factors such as temperature and pressure. Without the concept of time, we would not be able to determine how long it takes for the system to reach equilibrium or study the effect of temperature and pressure on the equilibrium position.

The concept of time also plays a critical role in the study of thermodynamics, which is the study of the transfer of energy between systems. Thermodynamics is a fundamental concept in chemistry as it explains the behaviour of chemical reactions under different conditions. The laws of thermodynamics describe the relationship between energy, work, heat, and temperature and provide a framework for understanding the behaviour of chemical systems. Without the concept of time, it would be challenging to study thermodynamics and understand the relationship between energy and matter.

The periodic table is a fundamental tool used in chemistry to organize and classify elements based on their properties. The arrangement of elements on the periodic table is dependent on their atomic number, which is a function of the number of protons in the nucleus. The concept of atomic number is dependent on the passage of time, as it is a function of the number of times an atom's nucleus has undergone a radioactive decay event.

Without the concept of time, it would be impossible to understand the behaviour of matter at the atomic and molecular level or study the transfer of energy between systems. Chemistry without the concept of time would be entirely different, and it is difficult to imagine how it would be. Time is an essential concept in science, and it provides a framework for understanding the natural world.

158

What is the origin of time?

Time is a fundamental concept in our understanding of the universe and the natural world. It is a concept that is deeply rooted in human perception, and it is difficult to imagine life without it. However, the origin of time is a complex question that has puzzled scientists and philosophers for centuries. In this article, we will explore some of the different theories that have been proposed to explain the origin of time.

The concept of time is intimately connected to the concept of change. Time can be seen as a way of measuring change, whether it is the change in the position of an object or the change in the state of a system. The origin of time, therefore, is intimately linked to the origin of change.

One of the earliest theories of the origin of time is the idea that time is eternal and infinite, with no beginning or end. This idea has roots in ancient Greek philosophy, with philosophers such as Aristotle and Parmenides arguing that time is an eternal and unchanging aspect of the universe. However, modern science has shown that this idea is not consistent with our current understanding of the universe. The Big Bang theory, which is the most widely accepted theory of the origin of the universe, suggests that the universe began with a singularity, a point of infinite density and temperature. The Big Bang is thought to have occurred around 13.8 billion years ago, and it marks the beginning of the universe as we know it today.

The concept of time, therefore, is intimately connected to the Big Bang. The universe, according to the Big Bang theory, began as a singularity, and time began at the moment of the Big Bang. This means that the origin of time is intimately linked to the origin of the universe.

Another theory of the origin of time is the idea that time is a human construct. According to this theory, time is not an objective aspect of the universe, but rather a subjective experience that is created by human perception. This idea has roots in philosophy, with philosophers such as Immanuel Kant arguing that time is a necessary condition for human perception.

However, this idea has been challenged by modern science. The laws of physics suggest that time is an objective aspect of the universe, and that it is intimately linked to the fabric of space-time. The theory of relativity, developed by Albert Einstein, suggests that space and time are intimately connected, and that the fabric of space-time can be warped by gravity. This means that time is not just a human construct, but rather an objective aspect of the universe that is intimately linked to the fabric of space-time.

Another theory of the origin of time is the idea that time is a consequence of entropy. Entropy is a measure of the disorder of a system, and it is intimately linked to the concept of time. According to this theory, time is a consequence of the increasing entropy of the universe. As the universe becomes more disordered over time, the concept of time emerges as a way of measuring this disorder.

This idea is consistent with our current understanding of the universe. The second law of thermodynamics, which is one of the fundamental laws of physics, suggests that entropy always increases over time. This means that the universe is moving towards a state of maximum entropy, where all matter and energy is evenly distributed throughout the universe. The concept of time, therefore, emerges as a consequence of the increasing entropy of the universe.

The Big Bang theory, which suggests that time began at the moment of the Big Bang. This theory is supported by a wealth of observational evidence, including the cosmic microwave background radiation, which is the afterglow of the Big Bang, and the observed expansion of the universe.

The question of what caused the Big Bang, and therefore the origin of time, is still a topic of much debate and research. Some theories suggest that the Big Bang was caused by a quantum fluctuation in a pre-existing universe, while others suggest that it was the result of a collision between two universes in a multiverse.

Another aspect of the origin of time that is worth considering is the arrow of time. The arrow of time refers to the fact that time appears to flow in a specific direction, from past to future. This directionality is intimately linked to the concept of entropy, as discussed earlier. The second law of thermodynamics suggests that entropy always increases over time, and this is what gives rise to the arrow of time.

The arrow of time is also intimately linked to the concept of causality. The idea of cause and effect is a fundamental aspect of our understanding of the universe, and it is intimately linked to the concept of time. Without the concept of time, it would be impossible to talk about cause and effect.

One interesting aspect of the origin of time that is worth considering is the philosophical and metaphysical implications of the concept. For example, the idea of a beginning to time raises questions about the nature of existence and the possibility of an ultimate cause or purpose to the universe.

We experience the world through a series of temporal events that occur in sequence, and our sense of self and identity is intimately linked to our experience of time. Without time, it is difficult to imagine what our experience of reality would be like, or even if we could have such an experience at all.

The origin of time is also closely linked to the concept of infinity. If time had a beginning, then it is natural to ask what came before that beginning. Similarly, if time has an end, what will happen after that end? These questions raise interesting philosophical and theological questions about the nature of infinity and the possibility of infinite regression.

It is worth considering the role of scientific inquiry in our understanding of the origin of time. Science has provided us with a powerful framework for understanding the physical world, but it is limited by the fact that it is based on empirical observation and testable hypotheses. The question of the origin of time, however, may lie beyond the reach of scientific inquiry, at least in its entirety. This raises interesting questions about the limits of scientific knowledge and the role of philosophy and theology in our understanding of the universe.

While we may never fully understand the nature of time or its origin, the pursuit of knowledge and understanding is a worthwhile endeavour that has the potential to expand our horizons and deepen our appreciation for the mysteries of the universe. The question of what caused the Big Bang, and therefore the origin of time, is still a topic of much debate and research, and it is an area of science that is sure to continue to fascinate and challenge us for many years to come.

THE PSYCHOLOGY OF TIME

TIME IS A FUNDAMENTAL aspect of our lives, yet our experience of time is far from straightforward. Time can seem to drag on endlessly or fly by in the blink of an eye, and our perception of time can vary depending on a range of factors, including our emotional state, level of attention, and even the cultural context in which we live. This article will explore the psychology of time, examining the ways in which we experience and perceive time, and the factors that influence our subjective experience of time.

The Perception of Time

Our perception of time is influenced by a variety of factors, including our level of arousal and attention, our emotional state, and the level of sensory stimulation we are experiencing. For example, when we are engaged in a task that requires our full attention, time can seem to pass more quickly, while when we are bored or disengaged, time can seem to drag on endlessly. Similarly, when we are experiencing a positive emotion, such as happiness or excitement, time can seem to fly by, while negative emotions, such as anxiety or depression, can make time seem to slow down.

The perception of time is also influenced by the cultural context in which we live. Different cultures have different attitudes towards time, with some cultures placing a high value on punctuality and efficiency, while others prioritize a more relaxed approach to timekeeping. These cultural differences can influence our perception of time and the level of importance we place on punctuality and timeliness.

The experience of time can also be influenced by our individual personality traits and cognitive styles. For example, individuals who are more impulsive or sensation-seeking may experience time as passing more quickly, while those who are more reflective or introverted may experience time as passing more slowly.

The Psychology of Time Perception

The psychology of time perception is a field of study that aims to understand the factors that influence our subjective experience of time. One key area of research in this field is the study of temporal illusions, which are situations in which our perception of time does not match up with objective time. For example, the time interval between two events can appear longer or shorter than it actually is, or time can seem to slow down or speed up in response to certain stimuli.

Temporal illusions can be induced by a variety of factors, including the level of sensory stimulation, the duration and intensity of the stimulus, and the level of attention and cognitive load required by the task. For example, the so-called "time flies when you're having fun" phenomenon can be explained by the fact that engaging in an enjoyable activity requires less cognitive effort and attention, leading to a faster subjective experience of time.

The psychology of time perception also encompasses the study of the neural and cognitive mechanisms that underlie our experience of time. Research in this area has identified a number of brain regions that are involved in the perception and processing of time, including the prefrontal cortex, the basal ganglia, and the parietal cortex.

THE RELATIONSHIP BETWEEN Time and Memory

Time is intimately linked to our ability to remember and perceive events in the past. Memory is essential for our ability to navigate the world and make decisions based on our past experiences, and the perception of time is a crucial aspect of memory encoding and retrieval.

Research has shown that our ability to remember events is influenced by the level of temporal detail in our memories. For example, memories that include more detailed temporal information, such as the sequence of events and the duration of each event, are more likely to be accurate and reliable than memories that lack this temporal detail.

The relationship between time and memory is also influenced by our emotional state. Emotionally charged events are more likely to be remembered vividly and in greater detail, and the level of emotional arousal experienced during an event can influence our perception of the duration of the event. For example, time can seem to slow down during a traumatic event, leading to the perception that the event lasted longer than it actually did.

The Relationship between Time and Motivation

The perception of time is also closely linked to our level of motivation and engagement with a task. When we are motivated and engaged, time can seem to pass more quickly, while when we are disengaged or unmotivated, time can seem to drag on. This phenomenon is known as "time dilation," and it occurs because our attentional focus is directed towards the task at hand, leading us to perceive time as passing more quickly.

Research has also shown that our perception of time can influence our level of motivation and goal-directed behaviour. For example, setting specific deadlines and time limits can increase motivation and productivity by providing a sense of urgency and focus.

THE PSYCHOLOGY OF FUTURE Time

The psychology of time also encompasses our perception and anticipation of future events. Our ability to plan for the future and make decisions based on future outcomes is crucial for our survival and well-being, and the perception of time plays a key role in this process.

Research has shown that our perception of future time is influenced by a range of factors, including our level of optimism, our past experiences, and our level of self-control. For example, individuals who are more optimistic tend to perceive future time as more open-ended and flexible, while those who are more pessimistic tend to perceive future time as more constrained and predetermined.

The perception of future time is also influenced by our ability to delay gratification and make decisions based on long-term outcomes. Research has shown that individuals who have high levels of self-control and are able to delay gratification tend to have a more positive outlook on the future and are more likely to make decisions that lead to long-term benefits.

Another interesting aspect of the psychology of time is the relationship between time and emotions. Our emotional state can have a significant impact on our perception of time, with positive emotions leading to an acceleration of time and negative emotions leading to a slowing down of time. For example, when we are engaged in an enjoyable activity, time can seem to fly by, while when we are experiencing anxiety or fear, time can seem to drag on.

The psychology of time also encompasses the study of time perception across different cultures and contexts. Different cultures have different beliefs and attitudes towards time, and these beliefs can influence how time is perceived and experienced. For example, cultures that place a greater emphasis on punctuality and efficiency may experience time as more rigid and structured, while cultures that value more relaxed and flexible attitudes towards time may experience time as more fluid and open-ended.

Finally, the psychology of time has important implications for our well-being and mental health. Dysfunctional time perception has been associated with a range of mental health conditions, including depression, anxiety, and post-traumatic stress disorder. Understanding the underlying mechanisms of time perception and its relationship with mental health can help us develop more effective interventions and therapies for these conditions.

Our perception of time is shaped by a variety of factors, including our emotional state, level of attention, cultural context, and cognitive style, and understanding the mechanisms that underlie our perception of time can provide valuable insights into human behaviour and decision-making. By continuing to explore the psychology of time, we can gain a deeper understanding of our experience of the world around us and the ways in which we navigate the complex and ever-changing landscape of time.

Our subjective experience of time is influenced by a variety of factors, including our emotional state, level of attention, cultural context, and cognitive style, and understanding the mechanisms that underlie our perception of time can provide valuable insights into human behaviour and decision-making. By exploring the psychology of time, we can gain a deeper understanding of our experience of the world around us and the ways in which we navigate the complex and ever-changing landscape of time.

How our perceptions of time influence our behaviour and decision-making?

Time is a fundamental aspect of human experience, and it plays a critical role in shaping our behaviour and decision-making. Our perception of time can influence how we allocate our resources, prioritize our goals, and evaluate the consequences of our actions. In this article, we will explore the ways in which our perceptions of time impact our behaviour and decision-making.

One way in which our perceptions of time influence our behaviour is by affecting our sense of urgency. When we perceive time as scarce, we are more likely to act quickly and decisively, often sacrificing long-term goals for short-term gains. This can lead to impulsive behaviour, as we prioritize immediate rewards over delayed gratification.

On the other hand, when we perceive time as abundant, we are more likely to take a measured approach, investing time and resources in long-term goals and delaying gratification for greater rewards in the future. This can lead to more deliberate decision-making, as we carefully weigh the costs and benefits of different courses of action.

Our perceptions of time can also influence our risk-taking behaviour. When we perceive time as short, we are more likely to take risks, as we feel a greater sense of urgency to achieve our goals before time runs out. This can lead to reckless behaviour, as we prioritize the potential rewards of our actions over the potential risks.

Conversely, when we perceive time as long, we are more likely to take a cautious approach, avoiding risks and seeking to minimize potential losses. This can lead to more conservative decision-making, as we focus on maintaining stability and security over the long-term.

Our perceptions of time can also influence our motivation and productivity. When we perceive time as limited, we are more likely to focus on urgent tasks, completing them quickly and efficiently to meet deadlines. However, this can come at the expense of longer-term goals, as we may neglect important tasks that have less immediate urgency.

On the other hand, when we perceive time as abundant, we are more likely to take a broader view, focusing on tasks that are important but less urgent. This can lead to greater creativity and innovation, as we have more time to explore new ideas and approaches.

Our perceptions of time can also influence our interpersonal relationships. When we perceive time as limited, we may feel a greater sense of urgency to accomplish tasks or spend time with loved ones. This can lead to stress and conflict if we feel that our time is being wasted or not being used effectively.

On the other hand, when we perceive time as abundant, we may feel more relaxed and open to spending time with others, even if it means delaying other tasks or goals. This can lead to deeper and more meaningful relationships, as we prioritize the importance of human connection and social support.

Moreover, our perceptions of time can also be influenced by cultural and societal factors. In some cultures, time is viewed as a valuable resource that should be used efficiently, while in others, time may be viewed as more fluid and less structured. These cultural differences can lead to misunderstandings and miscommunications between individuals from different cultural backgrounds.

Our perceptions of time can be influenced by external factors, such as technology and media. The constant availability of information and entertainment can create a sense of time pressure, as we feel the need to keep up with the latest news and trends. This can lead to distraction and decreased productivity, as we struggle to balance the demands of our work and personal lives.

By understanding the ways in which our perceptions of time influence our behaviour and decision-making, we can make more intentional choices about how we use our time, and cultivate a greater sense of purpose and fulfillment in our lives.

DEBASISH TALUKDAR

THE CULTURAL SIGNIFICANCE OF TIME

TIME IS A UNIVERSAL concept, but its meaning and significance can vary greatly across different cultures and societies. The way in which time is valued, measured, and perceived is deeply rooted in cultural beliefs and practices, and can play a significant role in shaping individual and collective identity.

In some cultures, time is viewed as a linear and progressive force, with a clear beginning and end. This view is often associated with a focus on productivity and efficiency, with individuals and societies striving to make the most of the limited time available to them. For example, in many Western societies, time is measured in precise units, such as seconds and minutes, and punctuality is highly valued as a sign of respect and professionalism.

In other cultures, time may be viewed in a more cyclical or circular manner, with a greater emphasis on the interconnectedness of past, present, and future. In these cultures, events and actions may be understood as repeating in a predictable cycle, rather than progressing linearly towards an endpoint. For example, in many indigenous cultures, time is measured in relation to natural cycles, such as the changing of the seasons or the movements of the stars.

The cultural significance of time is also reflected in the rituals and practices that are associated with it. For example, in many cultures, there are specific times of the day or year that are designated for religious or spiritual practices, such as prayer or meditation. In other cultures, specific times may be associated with social gatherings or celebrations, such as festivals or holidays.

The way in which time is valued and prioritized can also reflect broader cultural values and priorities. For example, in cultures that place a strong emphasis on family and community, time may be viewed as a resource to be shared and enjoyed collectively, rather than individually. In contrast, in cultures that prioritize individualism and self-reliance, time may be viewed as a personal resource to be managed and optimized for maximum productivity and efficiency.

Furthermore, the cultural significance of time can play a role in shaping individual and collective identity. Our relationship to time can influence our sense of purpose, motivation, and fulfilment, and can shape our perceptions of ourselves and our place in the world. For example, in cultures that prioritize long-term planning and goal-setting, individuals may be more likely to identify with their future selves and invest in activities that promote long-term well-being. In contrast, in cultures that prioritize present-moment experiences and pleasures, individuals may be more likely to identify with their immediate desires and pursue short-term gratification.

The cultural significance of time can also be seen in the ways in which it is represented in art, literature, and other forms of creative expression. For example, in literature, time can be used as a metaphor for change, growth, and the passage of life. Many famous works of literature, such as Charles Dickens' "A Christmas Carol" and Virginia Woolf's "Mrs. Dalloway," explore the themes of time and memory, and the impact that they have on our sense of self and identity.

In art, time can be represented in a variety of ways, such as through the use of clocks, calendars, or other symbols of the passage of time. For example, the painting "The Persistence of Memory" by Salvador Dali features melting clocks, which are a symbolic representation of the fluidity and subjectivity of time. Similarly, in the film "Boyhood," director Richard Linklater used the medium of film to capture the passage of time over a twelve-year period, allowing viewers to witness the gradual growth and development of the film's protagonist.

The cultural significance of time is also evident in the way that we use language to talk about time. Different languages have different structures and tenses that are used to describe the passage of time, and these structures can reflect broader cultural beliefs and values.

For example, in the Hopi language, there is no future tense, reflecting the culture's focus on living in the present moment and the interconnectedness of past, present, and future.

By exploring the ways in which time is valued, measured, and represented across different cultures and societies, we can gain a deeper understanding of the diversity and richness of human experience, and work towards building more inclusive and equitable communities.

How different culture's view and conceptualize time?

Time is a fundamental aspect of human existence, yet its conceptualization and measurement vary greatly across different cultures and societies. While the Western world tends to view time as a linear and measurable construct, other cultures have developed more cyclical or relational understandings of time. Let us try to explore some of the different ways in which cultures view and conceptualize time.

Linear Time

In Western cultures, time is often seen as a linear progression from past to present to future. This view is reflected in the use of clocks and calendars, which allow for precise measurement and planning of time. Western cultures also tend to prioritize punctuality and time management, valuing efficiency and productivity.

This linear view of time can be traced back to the ancient Greeks, who viewed time as a series of discrete moments that could be objectively measured and recorded. The development of clocks and other timekeeping devices during the Renaissance further solidified this view of time as a fixed and measurable quantity.

Linear time is closely associated with the idea of progress and achievement, with the belief that time moves inexorably forward towards a better future. This view of time has been closely linked to the rise of industrialization and capitalism, which emphasize the importance of efficiency and productivity in a linear timeline.

In terms of daily life, the concept of linear time has led to the development of rigid schedules and timelines, where events are carefully planned and managed to optimize productivity and achievement. This has led to a fast-paced and stressful lifestyle, where individuals are constantly working to keep up with the demands of their schedules.

Critics of linear time argue that this concept is limiting and narrow, and fails to take into account the cyclical nature of life and the interconnectedness of all things. They argue that linear time leads to a focus on individual achievement and progress, at the expense of community and the environment.

The linear view of time has also influenced the way people organize their lives and daily routines. Time is often seen as a scarce resource that must be used efficiently, leading to a fast-paced and hectic lifestyle. The pressure to be productive and make the most of every moment can lead to stress and burnout.

The linear view of time has also had a significant impact on social and political systems. For example, the idea of progress has been used to justify colonialism and the spread of Western culture, with the belief that Western societies are more advanced and civilized than others. This has led to the erasure of indigenous cultures and knowledge systems that view time in different ways.

Cyclical Time

In contrast to the linear view of time, many cultures view time as cyclical or repetitive. This view sees time as a series of cycles, such as the changing of seasons or the repeating patterns of birth, growth, and death. In cyclical time, events are seen as part of a larger pattern or rhythm, rather than isolated occurrences.

Many indigenous cultures, such as those of the Native American and Aboriginal Australian peoples, hold cyclical views of time. In these cultures, time is seen as a circle or spiral, with events and cycles repeating themselves over time. This cyclical view of time is often associated with a deep connection to nature and the natural world.

In cyclical time, events and seasons repeat in a circular manner, with no beginning or end. This concept of time is often linked to the cycles of nature, such as the changing of seasons or the cycles of the moon. In this view, time is seen as a part of a larger cosmic rhythm, and the cycles of time are seen as interconnected with the cycles of nature.

Cyclical time is often associated with the idea of rebirth and renewal. In many indigenous cultures, death is not seen as the end of life but rather a transformation or rebirth into a new form. This idea is reflected in many religious traditions, such as Hinduism and Buddhism, where the concept of reincarnation is central to the belief system.

Cyclical time also emphasizes the interconnectedness of all things. In this view, everything in the universe is part of a larger web of life, and events are seen as interconnected and interdependent. This understanding of time is reflected in many indigenous cultures where community and cooperation are emphasized over individualism and competition.

The cyclical view of time has had a significant impact on the way people organize their lives and daily routines. Time is often seen as abundant and flexible, allowing for a more relaxed and natural pace of life. The pressure to be productive and make the most of every moment is not as prevalent in cyclical time, as time is seen as a part of a larger cosmic rhythm that cannot be controlled.

Relational Time

In some cultures, time is seen as a relational construct, with events and experiences taking on different meanings depending on their relationship to other events and experiences. In these cultures, time is less about objective measurement and more about the subjective experiences of individuals and communities.

Unlike linear and cyclical time, which see time as an objective and measurable quantity, relational time sees time as a social construct that is shaped by our relationships and experiences.

In relational time, the way we experience time is influenced by our social and cultural context, as well as by our individual perspectives and beliefs. Time is not seen as an external force that is independent of human experience, but rather as a product of our relationships with others and our environment.

One of the key aspects of relational time is the idea that time is experienced differently by different people. This is because our experiences of time are shaped by factors such as our culture, gender, age, and social class. For example, the experience of time may be different for someone living in a fast-paced urban environment compared to someone living in a more rural setting. Similarly, the experience of time may be different for someone who is retired compared to someone who is in the midst of a demanding career.

Relational time also emphasizes the importance of context in shaping our experiences of time. Time is not seen as a fixed quantity, but rather as something that is shaped by our interactions with the world around us. For example, the experience of time may be different when we are engaged in a meaningful activity compared to when we are engaged in a mundane task. Similarly, the experience of time may be different when we are with people we love compared to when we are with people we don't know well.

Another important aspect of relational time is the idea that time is not simply a passive backdrop to human experience, but rather an active force that shapes our relationships and experiences. For example, the experience of time may influence the way we relate to others and the way we understand ourselves. The way we perceive time can also influence the way we make decisions and the way we prioritize our goals and values.

Unlike linear time, which sees time as a fixed quantity that progresses uniformly and independently of human experience, relational time recognizes that the experience of time is highly variable and influenced by our interactions with others and our environment.

Relational time also highlights the role of human agency in shaping our experience of time. While time may be influenced by a range of social and cultural factors, we also have the power to shape our own experience of time. By being mindful of our relationship with time and the way we prioritize our goals and values, we can shape our experience of time in a way that feels more fulfilling and meaningful.

Spatial Time

Another way in which cultures conceptualize time is through space. In some cultures, time is seen as a spatial concept, with events and experiences located in different points in space. This view is reflected in the use of spatial metaphors to describe time, such as "looking back" on past events or "moving forward" into the future.

In Chinese culture, time is often represented as a circle or wheel, with events located at different points around the circle. This view of time emphasizes the cyclical nature of time, while also acknowledging the importance of space and location.

Spatial time is a relatively new concept that has emerged from the intersection of physics and philosophy. It is based on the idea that time and space are fundamentally intertwined, and that our experience of time is inseparable from our experience of space.

In traditional physics, time is treated as a separate dimension from space, with time being the fourth dimension that complements the three dimensions of space. However, recent advances in physics, particularly in the field of relativity, have challenged this traditional view. According to the theory of relativity, time and space are not separate dimensions, but rather different aspects of a single four-dimensional entity known as spacetime.

From a philosophical perspective, spatial time is based on the idea that our experience of time is intimately connected to our experience of space. This is because our experience of time is always situated within a particular space or environment. For example, our experience of time may feel very different in a crowded, noisy city environment compared to a peaceful natural setting.

One of the key features of spatial time is that it emphasizes the importance of context in shaping our experience of time. This means that the way we experience time is influenced by the physical, social, and cultural environment in which we find ourselves. For example, the pace of life in a busy city may create a sense of time passing more quickly, while a slower, more relaxed environment may create a sense of time passing more slowly.

Spatial time also recognizes that our experience of time is not uniform, but rather varies depending on our position within space. This is because time can be influenced by factors such as gravity and velocity. According to the theory of relativity, time passes more slowly in environments with higher gravity or at higher velocities, meaning that our experience of time can differ depending on our position within space.

Another important aspect of spatial time is the idea that time and space are not passive backdrops to human experience, but rather active forces that shape our perceptions and behaviour. The way we experience time and space can influence the decisions we make, the relationships we form, and the way we interact with our environment.

The role of time in religious and spiritual practices

Time plays a significant role in religious and spiritual practices across cultures and traditions. These practices are often built around the idea of time as a cycle, an endless repetition of events that can be controlled and shaped through ritual and observance. The idea of time in religious practices is not just a measure of time but is also a measure of the sacred.

Time is an important concept in many religions and spiritual practices. Here is a brief overview of how time is viewed and valued in some of the major world religions:

CHRISTIANITY:

Christians view time as a finite and linear progression, with a definite beginning and end. The belief in a linear, finite time is a fundamental part of the Christian worldview, and it is based on the belief that God created the world and that history has a predetermined end.

The Bible describes time as having a beginning and an end, with a specific purpose for human life in between. Christians believe that God created the world in six days and rested on the seventh, establishing the concept of the seven-day week. The concept of time is closely tied to the concept of salvation in Christianity, as time is seen as a finite opportunity for humans to accept God's offer of salvation and achieve eternal life.

The Christian calendar is also an important aspect of the role of time in Christianity. The liturgical calendar is based on the life of Jesus Christ and marks the events of his life, such as his birth, death, and resurrection. The calendar provides a structured way for Christians to observe and commemorate these events, as well as other important religious observances and feasts.

Christian eschatology, or the study of the end of the world, is also closely tied to the role of time in Christianity. The Bible predicts the end of the world and the return of Jesus Christ, which Christians believe will happen at an unknown time. Christians believe that the end of the world is a predetermined event and that it will happen according to God's plan.

In addition, the role of time in Christianity is evident in the practices of daily prayer, weekly worship, and annual religious observances. Christians believe that these practices help them to stay connected to God and to experience spiritual growth and renewal.

The Christian view of time emphasizes the importance of living in the present moment, making the most of the time we have, and being mindful of God's plan for our lives.

ISLAM:

Muslims believe that time is a precious resource and a gift from God, which should be used wisely to worship God, do good deeds, and seek knowledge.

In Islam, time is viewed as a continuous cycle of creation and destruction, which will ultimately come to an end on the Day of Judgment. Muslims believe that God created the world and all of its inhabitants and that everything that happens in life is predetermined by God. This belief in predestination leads Muslims to value the present moment and focus on what they can do now to please God.

The Islamic calendar is based on the lunar cycle and consists of twelve months, with each month corresponding to a particular event or observation. The most significant event on the Islamic calendar is the month of Ramadan, during which Muslims fast from dawn until dusk and engage in increased worship and charity. Ramadan is considered a time of spiritual renewal and self-reflection, during which Muslims seek to purify themselves and increase their devotion to God.

In addition to the calendar, time is also significant in Islamic prayer. Muslims are required to pray five times a day, and these prayers are performed at specific times throughout the day. These prayers are seen as a way to connect with God and maintain a spiritual connection throughout the day.

Muslims also believe in the concept of the Hereafter, a time when all individuals will be judged according to their deeds and either rewarded or punished in the afterlife. This belief encourages Muslims to live their lives in a way that is pleasing to God and to use their time on Earth to perform good deeds and seek forgiveness for their sins.

Time is seen as a gift from God, and Muslims are encouraged to make the most of it by engaging in acts of worship, performing good deeds, and seeking knowledge. The Islamic calendar and prayer times provide a structured way for Muslims to organize their time and prioritize their spiritual obligations, ultimately leading to a life of purpose and fulfilment.

Hinduism:

Hinduism is one of the oldest and most complex religions in the world. Time plays a significant role in Hinduism, and it is viewed from both a cyclical and linear perspective. Hinduism has a unique way of conceptualizing time, which is different from other major religions. The concept of time in Hinduism is closely related to the concepts of karma, reincarnation, and the cycle of birth and death.

In Hinduism, time is divided into cycles of creation and destruction. This cycle is known as the Kalachakra, and it is said to repeat itself endlessly. Each cycle consists of four yugas or ages: Satya Yuga, Treta Yuga, Dwapara Yuga, and Kali Yuga. Satya Yuga is considered the golden age, where people lived in harmony with nature and followed dharma (righteousness) in all aspects of life. As we move towards Kali Yuga, the age of darkness and ignorance, people become more selfish, and morality declines.

The Hindu concept of time is also closely related to the concept of karma. Karma refers to the law of cause and effect, where every action has a consequence. The actions we take in our current life will affect our future lives, either positively or negatively. Therefore, the present moment is considered crucial, and it is believed that the actions we take in this life will determine our future.

Another important aspect of time in Hinduism is the celebration of festivals and rituals. Hindus have a rich tradition of celebrating festivals that mark important events in their history and mythology. Many of these festivals are based on the lunar calendar and are celebrated on specific dates each year. These festivals are an essential part of Hindu culture, and they help to bring people together and reinforce their sense of community.

In Hinduism, time is also considered an illusion or maya. It is believed that the ultimate reality is eternal and beyond time and space. The ultimate goal of life in Hinduism is to realize this ultimate reality and to break free from the cycle of birth and death.

Hinduism divides time into four ages, or "yugas." The first yuga is the "Satya Yuga," also known as the "Golden Age." This yuga is said to be the most prosperous and virtuous of all the ages. The second yuga is the "Treta Yuga," followed by the "Dvapara Yuga" and the "Kali Yuga," which is the current age we are living in.

Hinduism also has a complex system of measuring time, known as the "cosmic cycle of time," which is made up of smaller cycles. One of the most important cycles is the "yuga cycle," which is a cycle of four ages that repeats itself endlessly. Each yuga has a different duration and quality, with the Satya Yuga lasting for 1,728,000 years and the Kali Yuga lasting for 432,000 years.

Time is also a crucial aspect of Hindu worship, with many rituals and prayers being performed at specific times of the day or year. For example, the daily worship of the gods and goddesses, known as "puja," is typically performed at sunrise and sunset. The celebration of major festivals such as Diwali, Navratri, and Dussehra also occurs at specific times of the year, according to the lunar calendar.

time plays a crucial role in Hinduism, both in its cosmic and individual manifestations. It is seen as a force that governs all aspects of life, from the cycles of the universe to the cycles of birth and death. Understanding the nature of time is considered essential for spiritual progress and the attainment of liberation.

BUDDHISM:

In Buddhism, time is viewed as a constantly changing and impermanent aspect of existence. The concept of time in Buddhism is closely related to the concept of impermanence, which states that all things are subject to change and are in a constant state of flux.

Buddhism recognizes two types of time: conventional time and ultimate time. Conventional time is the time that we experience in our daily lives, with past, present, and future. Ultimate time, on the other hand, is the timeless and eternal nature of reality that exists beyond our conventional experience.

The Buddhist teachings emphasize the importance of being mindful and present in the current moment, as it is the only moment that truly exists. The past and future are seen as mental constructs that can distract us from the present moment and cause us suffering.

In Buddhism, the cyclical nature of time is also recognized, with the concept of Samsara, the cycle of birth, death, and rebirth. The cycle is believed to be driven by karma, the law of cause and effect, which determines the quality of a person's life and their next rebirth.

Buddhism also recognizes the impermanence of all things, including time itself. The teachings emphasize that everything is in a constant state of change and that clinging to things that are impermanent can cause suffering.

In terms of religious practices, time plays a significant role in Buddhist festivals and rituals, such as Vesak (Buddha's birthday), Asalha Puja (the day Buddha gave his first sermon), and Kathina (a ceremony for offering robes to monks). These celebrations are typically held on specific lunar calendar dates and involve practices such as meditation, chanting, and giving offerings.

JUDAISM:

Jews believe in the existence of a single, omnipotent God who created the universe and everything in it. According to Jewish tradition, God created the world in six days and rested on the seventh day, which is observed as the Sabbath.

The concept of time in Judaism is based on a lunar calendar, which is different from the Gregorian calendar that is commonly used in the Western world. The Jewish calendar consists of twelve lunar months, and each month is either 29 or 30 days long. This means that the Jewish calendar year is shorter than the Gregorian calendar year, which has 365 or 366 days.

One of the most important religious holidays in Judaism is the celebration of the Jewish New Year, which is known as Rosh Hashanah. This holiday marks the beginning of the Jewish calendar year and is usually observed in September or October. During Rosh Hashanah, Jews reflect on the past year and pray for a good and fruitful year ahead.

Another significant holiday in Judaism is Yom Kippur, also known as the Day of Atonement. This holiday occurs ten days after Rosh

Hashanah and is considered the holiest day of the year in Judaism. On Yom Kippur, Jews fast for twenty-four hours and spend the day in prayer and reflection, asking for forgiveness for their sins.

The concept of time in Judaism is also reflected in the idea of the Messiah, who is believed to be a saviour who will come to redeem the Jewish people and bring about the end of the world as we know it. According to Jewish tradition, the coming of the Messiah is imminent, but the timing is unknown.

SIKHISM:

Sikhism, founded by Guru Nanak in the 15th century, emphasizes the importance of the present moment and living in harmony with the natural flow of time. Time is seen as a gift from God and an opportunity to connect with the divine.

One of the central tenets of Sikhism is the concept of "Hukam," which means divine order or will. Sikhs believe that everything in the universe happens according to God's will and that everything is part of a greater plan. As a result, time is viewed as a manifestation of God's will, and every moment is considered sacred.

Sikhs mark the passage of time with a variety of religious festivals and observances, many of which are based on the lunar calendar. The most important festival is Gurpurab, which celebrates the birth anniversaries of the ten Sikh gurus. Other important festivals include Vaisakhi, which marks the founding of the Khalsa, and Diwali, which celebrates the triumph of light over darkness.

Sikhs also place a great deal of importance on daily prayer and meditation as a way to connect with the divine and live in harmony with the natural flow of time. They believe that by focusing on the present moment and cultivating a sense of inner peace and contentment, they can experience a deeper connection with the divine and a greater sense of purpose in life.

INDIGENOUS RELIGIONS:

Many indigenous traditions share a deep respect for the natural world and an understanding that time is interconnected with the cycles and rhythms of nature.

For example, many indigenous cultures mark the passage of time with seasonal ceremonies and rituals that honour the changing of the seasons and the natural cycles of growth and decay. These ceremonies often involve offerings to the spirits or deities associated with the land, water, and animals, and are intended to ensure a bountiful harvest and a healthy ecosystem.

In many indigenous cultures, time is also viewed as a cyclical rather than linear process, with events repeating themselves in a never-ending cycle. This cyclical view of time can be seen in the use of creation stories and mythologies that describe the origins of the world and the cycles of birth, death, and rebirth.

In some indigenous traditions, time is viewed as a multi-dimensional phenomenon, with different layers of reality existing simultaneously. For example, the Navajo tradition of "hózhó" emphasizes the importance of living in harmony with all aspects of creation, including the past, present, and future. In this worldview, time is seen as a continuum rather than a linear progression, and events in one time period can influence events in another.

The impact of globalization on our perception of time

Globalization has brought about significant changes in our perception of time. As the world becomes more interconnected, different cultures and traditions have come into contact with one another, leading to a blending of perspectives and a reshaping of ideas about time.

One of the major impacts of globalization on our perception of time is the acceleration of time. Advances in technology and communication have led to an increase in the speed at which we can connect and exchange information with one another. As a result, our lives have become more fast-paced, and there is a greater emphasis on efficiency and productivity. This acceleration of time can create a sense of pressure to constantly be doing more and achieving more, leading to feelings of stress and burnout.

Another impact of globalization on our perception of time is the breakdown of traditional temporal boundaries. As people from different cultures and time zones interact more frequently, the boundaries between different temporal norms and customs become blurred. For example, in the globalized world, it is increasingly common to work with colleagues or clients in different time zones, which requires a flexible and adaptable approach to time management.

The impact of globalization on our perception of time is also reflected in the rise of global cultural trends. Globalization has led to the spread of popular culture and fashion trends around the world, which often have a temporal dimension. For example, the fashion industry operates on a seasonal cycle, with new collections being released every few months. This trend-driven approach to fashion has led to a culture of disposability, where clothing and accessories are quickly discarded and replaced with the latest trend.

Globalization has also created a sense of temporal dislocation, as people are increasingly connected to events and issues in different parts of the world. For example, people can watch news events happening on the other side of the world in real-time, creating a sense of connection and immediacy to events that would have been unimaginable in the past.

The impact of globalization on our perception of time is complex and multifaceted, and it is likely to continue to evolve in the years to come. While globalization has created new challenges and pressures, it has also opened up new opportunities for cross-cultural understanding and exchange. As we continue to navigate the challenges and opportunities of the globalized world, it is important to remain aware of the ways in which our perceptions of time are changing, and to develop strategies for managing time in a way that is healthy and sustainable.

With the advancement of technology and transportation, the world has become more connected, and the pace of life has accelerated. The concept of time has become more standardized and synchronized across different parts of the world.

One of the significant impacts of globalization on our perception of time is the standardization of time zones. Before the advent of globalization, each region had its own method of measuring time, based on the position of the sun. However, with the growth of international trade and transportation, it became necessary to standardize time. In 1884, the International Meridian Conference in Washington, D.C. established a system of 24 standard time zones, which became the basis of the modern time zone system.

Globalization has also led to the adoption of a more linear concept of time. The modern business world operates on a tight schedule, with deadlines and time-sensitive projects. This emphasis on punctuality and efficiency has led to a perception of time as a valuable commodity that needs to be managed and optimized. This linear concept of time is in contrast to cyclical and relational concepts of time that are prevalent in many traditional societies.

Globalization has also led to the increasing homogenization of cultures and lifestyles. The spread of Western cultural values and practices has influenced the way people perceive time in different parts of the world. The pace of life has accelerated, and people are expected to be constantly connected and available, regardless of their location or time zone.

The impact of globalization on our perception of time is not entirely negative, however. The standardization of time zones and the emphasis on punctuality and efficiency have led to greater efficiency in business and communication. It has also made it easier for people to travel and work in different parts of the world.

As we continue to navigate the complex terrain of globalization, it is important to be mindful of the impact it has on our perception of time and the values and beliefs that shape our relationship with time.

THE FUTURE OF TIME

The concept of time has fascinated humans for centuries, and as we continue to progress technologically and intellectually, the future of time remains an intriguing topic of discussion. Many people wonder how our understanding of time will evolve and how it will affect our lives. In this article, we will explore some possible future scenarios for the concept of time.

One possible future for time is that it will become even more malleable than it already is. We have already seen how time can be manipulated with the use of technologies like GPS and atomic clocks, but the future could bring even greater advancements in our ability to control time. This could include the creation of new technologies that allow us to slow down or speed up time, or even the ability to manipulate time on a quantum level.

Another possibility is that our understanding of time will become more complex as we continue to explore the mysteries of the universe. For example, physicists are already studying the concept of time travel and the possibility of multiple universes, which could greatly impact our perception of time. As we learn more about these concepts, we may begin to see time in new ways that we never thought possible.

Our perception of time could be greatly impacted by cultural changes and globalization. As we continue to become more interconnected as a global society, we may see a blending of different cultural perceptions of time. This could lead to a greater understanding and appreciation of different approaches to time, or it could lead to a homogenization of time that erases cultural differences.

The future of time also raises questions about the impact it will have on our daily lives. Will we become even more obsessed with time, or will we learn to embrace a more flexible approach to our schedules? Will time become an even greater source of stress and anxiety, or will we find ways to live in the moment and appreciate the present?

One thing is certain: time will continue to play a central role in our lives, and as we continue to evolve and progress as a society, our understanding and perception of time will evolve with us. While we cannot predict exactly what the future of time will look like, we can be sure that it will continue to fascinate and intrigue us for generations to come.

As we continue to advance technologically, our understanding and experience of time will inevitably change. One possible future scenario is that we will become even more obsessed with efficiency and productivity, as the pressure to maximize output and minimize wasted time grows. This could lead to an even more fragmented and accelerated experience of time, as we try to cram more and more activities into each moment.

Alternatively, we may begin to realize the limitations of this approach, and seek out ways to slow down and savour our experiences. This could involve a renewed appreciation for activities that are inherently slow-paced, like gardening, cooking, or reading. We may also see a resurgence of interest in mindfulness practices and other forms of meditation, which encourage us to be fully present in the moment and to let go of our constant need to multitask.

But one thing is certain: time will continue to be a fundamental aspect of our lives, shaping the way we think, feel, and interact with the world around us. As we move forward into an uncertain future, it will be important to remain mindful of our relationship with time, and to strive to cultivate a healthy and balanced approach to this ever-present aspect of our existence.

Advances in technology and the impact on our relationship with time

Advances in technology have had a significant impact on the way we perceive and interact with time. From the invention of the clock to the emergence of the internet and social media, technology has shaped our relationship with time in numerous ways.

The clock was one of the most significant technological advances in human history. Before the invention of the clock, time was measured using natural phenomena such as the movement of the sun or the changing of seasons. However, with the invention of the clock, time became standardized and could be measured accurately, leading to a greater level of efficiency and productivity in society.

The rise of smartphones, the internet, and other forms of technology have transformed the way we think about and experience time. Today, we are constantly connected to the world around us, and we are able to access information and communicate with others at any time of the day or night.

One of the key ways in which technology has impacted our relationship with time is through the acceleration of our daily lives. We are able to complete tasks faster than ever before, and we expect instant gratification in all aspects of our lives. We can order food, clothes, and other goods online and have them delivered to our doorsteps within hours. We can communicate with people on the other side of the world in real-time, and we can access information on virtually any topic with just a few clicks.

At the same time, technology has also allowed us to become more efficient with our time. We can automate tasks that would have taken hours or even days to complete in the past. We can use software to manage our schedules and prioritize our tasks, which allows us to be more productive and accomplish more in less time.

However, there are also downsides to the impact of technology on our relationship with time. The constant barrage of notifications, emails, and other distractions can lead to a feeling of being overwhelmed and stressed. The pressure to respond to messages and complete tasks quickly can also lead to a sense of urgency and anxiety.

The constant availability of technology can blur the boundaries between work and personal time, leading to an "always on" mentality that can be difficult to escape. This can lead to feelings of burnout and a sense of never truly being able to disconnect and recharge.

Additionally, the rise of artificial intelligence and automation has the potential to further transform our relationship with time. As machines become increasingly capable of performing tasks traditionally done by humans, the concept of time may become even more abstract. If machines can work around the clock without getting tired or needing breaks, how will this impact our understanding of work and leisure time?

As technology continues to evolve, it will be important to consider the impact on our perception and use of time.

CONCLUSION

Time has been a fascinating subject of inquiry across various disciplines, including physics, philosophy, psychology, anthropology, and religion. The book of time highlights the different ways in which we experience time, the cultural significance of time, and the role of time in religious and spiritual practices.

We have seen how time is viewed differently in different cultures, ranging from the cyclical time of Hinduism to the linear time of Western cultures. We have also explored the impact of globalization and technology on our relationship with time, with the increasing pace of life and the pressure to be productive.

The possibility of time travel, though still a topic of debate and speculation, has been explored by scientists and writers alike, with various theories and ideas put forward. While the practicality and feasibility of time travel remain uncertain, the concept has captured the human imagination and continues to inspire creativity and curiosity.

The future of time is likely to see further advances in technology and scientific discoveries, which will undoubtedly shape our understanding and relationship with time. As we continue to navigate the complexities of time, it is essential to recognize its importance and value, to use it wisely and cherish every moment we have.

Throughout history, humans have attempted to measure and control time through the development of calendars, clocks, and other technologies. These innovations have allowed us to track the passage of time more accurately and efficiently, but they have also had significant cultural and social impacts.

As we continue to advance technologically and explore the mysteries of the universe, the concept of time remains a fascinating and ever-evolving subject. We may never fully understand the true nature of time, but our continued exploration and understanding of this concept will undoubtedly lead to new insights and discoveries.

Whether we are seeking to understand the origin of the universe, the role of time in cultural practices, or the impact of technology on our lives, time will remain a fascinating and essential topic that will continue to inspire exploration and inquiry in the years to come.

Final thoughts on the significance of time and its role in shaping our world and our lives

Time is an abstract concept that permeates every aspect of our lives. It governs our every action, influences our behaviour and decision-making, and shapes our relationships with each other and with the world around us. In this book, we have explored the many dimensions of time, delving into its scientific, cultural, and philosophical aspects, and examining how it has been perceived and experienced across various cultures and periods of history.

We have seen that time is a complex and multifaceted phenomenon that defies easy categorization. From the linear time of modern Western societies to the cyclical time of traditional indigenous cultures, time is experienced and understood in different ways by different people. Yet despite these differences, time remains a fundamental part of the human experience, shaping our perceptions of the world and influencing our very sense of self.

One of the key themes that has emerged from our exploration of time is the idea that time is not simply a passive backdrop to our lives, but an active force that shapes our experiences and behaviours. Our perceptions of time can influence the way we approach tasks and make decisions, and can even affect our physical and mental well-being. Moreover, time is intimately connected to the natural world and the laws of physics, playing a central role in the functioning of our universe.

Another key theme that has emerged is the impact of globalization and technological advances on our relationship with time. As the world becomes more interconnected and our lives become more digitized, our perception of time has undergone significant changes. The rise of social media and instant messaging has accelerated the pace of life, blurring the boundaries between work and leisure and creating a culture of constant connectivity. At the same time, technological innovations like smartphones and GPS have transformed our understanding of time and space, allowing us to navigate the world in new ways and to access information and services on demand.

As we look to the future, it is clear that time will continue to play a central role in our lives and in the world around us. From the ongoing quest for a unified theory of time to the potential for time travel and other technological innovations, the study of time remains a fascinating and vital area of inquiry. Moreover, as the world faces new challenges and opportunities, from climate change to geopolitical conflict, our ability to understand and navigate the complexities of time will be essential to our ability to thrive and flourish.

By gaining a deeper understanding of time and its many dimensions, we can gain new insights into our world and ourselves, and can learn to navigate the challenges and opportunities that lie ahead. Whether we are contemplating the mysteries of the universe or simply reflecting on our own lives, the study of time offers a fascinating and endlessly rewarding journey of discovery.